AF412889

JEALOTT'S MILL
LIBRARY

ORGANIC SYNTHESES

Withdrawn from syngenta Library Centre

ORGANIC SYNTHESES

AN ANNUAL PUBLICATION OF SATISFACTORY
METHODS FOR THE PREPARATION
OF ORGANIC CHEMICALS

VOLUME 53

1973

ADVISORY BOARD

C. F. H. ALLEN

RICHARD T. ARNOLD

HENRY E. BAUMGARTEN

A. H. BLATT

VIRGIL BOEKELHEIDE

RONALD BRESLOW

T. L. CAIRNS

JAMES CASON

J. B. CONANT

E. J. COREY

WILLIAM G. DAUBEN

WILLIAM D. EMMONS

ALBERT ESCHENMOSER

L. F. FIESER

R. C. FUSON

HENRY GILMAN

C. S. HAMILTON

W. W. HARTMAN

E. C. HORNING

JOHN R. JOHNSON

WILLIAM S. JOHNSON

N. J. LEONARD

B. C. McKUSICK

C. S. MARVEL

MELVIN S. NEWMAN

C. R. NOLLER

W. E. PARHAM

CHARLES C. PRICE

NORMAN RABJOHN

JOHN D. ROBERTS

R. S. SCHREIBER

JOHN C. SHEEHAN

RALPH L. SHRINER

H. R. SNYDER

MAX TISHLER

KENNETH B. WIBERG

PETER YATES

BOARD OF EDITORS

ARNOLD BROSSI, *Editor-in-Chief*

RICHARD E. BENSON

GEORGE H. BÜCHI

HERBERT O. HOUSE

ROBERT E. IRELAND

CARL R. JOHNSON

SATORU MASAMUNE

WATARU NAGATA

ZDENEK VALENTA

WAYLAND E. NOLAND, *Secretary to the Board*
University of Minnesota, Minneapolis, Minnesota

FORMER MEMBERS OF THE BOARD, NOW DECEASED

ROGER ADAMS

HOMER ADKINS

WERNER E. BACHMANN

WALLACE H. CAROTHERS

H. T. CLARKE

ARTHUR C. COPE

NATHAN L. DRAKE

OLIVER KAMM

LEE IRVIN SMITH

FRANK C. WHITMORE

JOHN WILEY AND SONS
NEW YORK · LONDON · SYDNEY · TORONTO

Copyright © 1973, by John Wiley & Sons, Inc.

All rights reserved. Published simultaneously in Canada.

No part of this book may be reproduced by any means, nor transmitted, nor translated into a machine language without the written permission of the publisher.

"John Wiley & Sons, Inc. is pleased to publish this volume of Organic Syntheses on behalf of Organic Syntheses, Inc. Although Organic Syntheses, Inc. has assured us that each preparation contained in this volume has been checked in an independent laboratory and that any hazards that were uncovered are clearly set forth in the write-up of each preparation, John Wiley & Sons, Inc. does not warrant the preparations against any safety hazards and assumes no liability with respect to the use of the preparations."

Library of Congress Catalog Card Number: 21-17747

ISBN 0-471-10615-1

Printed in the United States of America

10 9 8 7 6 5 4 3 2 1

HANS THACHER CLARKE   LEE IRVIN SMITH

(1887–1972)   (1891–1973)

We regretfully note the deaths of Hans Thacher Clarke and Lee Irvin Smith. These chemists, pioneers in their specialties, contributed greatly to the development of *Organic Syntheses*. Hans Clarke edited Volumes 3 and 10 and Lee Smith edited Volumes 22 and 23. They were esteemed colleagues—we shall miss both of them.

The Active Board
The Advisory Board

NOMENCLATURE

Preparations appear in the alphabetical order of common names of the compounds or names of the synthetic procedures. For convenience in surveying the literature concerning any preparation through *Chemical Abstracts* subject indexes, the *Chemical Abstracts* indexing name for each compound is given as a subtitle if it differs from the common name used as the title.

SUBMISSION OF PREPARATIONS

Chemists are invited to submit for publication in *Organic Syntheses* procedures for the preparation of compounds that are of general interest, as well as procedures that illustrate synthetic methods of general utility. It is fundamental to the usefulness of *Organic Syntheses* that submitted procedures represent optimum conditions, and the procedures should have been checked carefully by the submitters, not only for yield and physical properties of the products, but also for any hazards that may be involved. Full details of all manipulations should be described, and the range of yield should be reported rather than the maximum yield obtainable by an operator who has had considerable experience with the preparation. For each solid product the melting-point range should be reported, and for each liquid product the range of boiling point and refractive index should be included. In most instances, it is desirable to include additional physical properties of the product, such as ultraviolet, infrared, mass, or nuclear magnetic resonance spectra, and criteria of purity such as gas chromatographic data. In the event that any of the reactants are not readily commerically available at reasonable cost, their preparation should be described in as complete detail and in the same manner as the preparation of the product of major interest. The sources of the reactants should be described in notes, and the physical properties (such as boiling point, index of refraction, melting point) of the reactants should be included except where standard commercial grades are specified.

Beginning with Volume 49, Sec. 3., Method of Preparation, and Sec. 4., Merits of the Preparation, have been combined into a single new Sec. 3., Discussion. In this section should be described other practical methods for accomplishing the purpose of the procedure that have appeared in the literature. It is unnecessary to mention methods that have been published but are of no practical synthetic value. Those features of the procedure that recommend it for publication in *Organic Syntheses* should be cited (synthetic method of considerable scope, specific compound of interest not likely to be made available commercially, method that gives better yield or is less laborious than other methods, etc.). If possible, a brief discussion of the scope and limitations of the procedure as applied to other examples as well as a comparison of the method with the other methods cited should be included. If necessary to the understanding or use of the method for related syntheses, a brief discussion of the mechanism may be placed in this section. The present emphasis of *Organic Syntheses* is on model procedures rather than on specific compounds (although the latter are still welcomed), and the Discussion section should be written to help the reader decide whether and how to use the procedure in his own research. Three copies of each procedure should be submitted to the Secretary of the Editorial Board. It is sometimes helpful to the Board if there is an accompanying letter setting forth the features of the preparations that are of interest.

Additions, corrections, and improvements to the preparations previously published are welcomed and should be directed to the Secretary.

PREFACE

This volume continues the policy of its predecessors by emphasizing the preparation of new and versatile chemicals as well as model procedures that are of general interest. Positive identification of intermediates and end products is made by giving pertinent spectroscopic information. At the same time hazardous reactions and exposure to dangerous chemicals are mentioned prominently.

Three procedures, which involve organometallic reactions, broaden the utility of this increasingly important type of reaction. Thus the homogeneous catalytic hydrogenation of DIHYDROCARVONE, the oxymercuration-reduction to give 1-METHYLCYCLOHEXANOL, and a variation of the well-known Clemmensen reduction, providing conditions that allow selective deoxygenation of ketones in polyfunctional molecules, are mentioned. Syntheses using organoboranes are illustrated by the preparation of ketones and alcohols via borane intermediates, by the introduction of sodium cyanoborohydride in hexamethylphosphoramide as a potent and convenient reagent for the reduction of halides and tosylates to the corresponding hydrocarbons, and by the use of diborane for reducing dinitriles. The use of lithium aluminum tri-*tert*-butoxyhydride for the reduction of acid chlorides to aldehydes is illustrated with the preparation of 3,5-DINITROBENZALDEHYDE.

Other new reagents of preparative value are involved in the selective bromination of an aralkyl ketone with phenyltrimethylammonium tribromide, the partial ether cleavage of aromatic *O*-methyl ethers with sodium ethyl mercaptide in dimethylformamide, and the preparation of DIDEUTERIODIAZOMETHANE which could be an important reagent for labeling studies. The preparation of furans using acetylenic sulfonium salts, the direct lithiation of aromatic compounds, and the preparation of 4,4-disubstituted cyclohexen-2-ones by addition

of methyl vinyl ketones to enamines—an interesting variation
of the Robinson annelation reaction—are properly described
with appropriate model examples. This group of new methods
of general interest is completed by discussions of the isoxazole
ring annelation, illustrated by the preparation of 1-METHYL-
4,4a,5,6,7,8-HEXAHYDRONAPHTHALEN-2(3*H*)-ONE; a
useful method for preparing β-hydroxyesters from esters and
ketones with the help of lithium bis(trimethylsilyl)amide; the
preparation of α,β-unsaturated aldehydes using diethyl 2-(cyclohexylamino)vinylphosphate via a Wittig reaction; the stereoselective synthesis of trisubstituted olefins; and the base-induced
rearrangement of epoxides illustrated by the preparation of
trans-PINOCARVEOL.

Finally, the syntheses of specific compounds include those
of ISOCROTONIC ACID, 1-PHENYL-4-PHOSPHORINAN-
ONE, ADAMANTANONE, AZETIDINE, DIAMANTANE
and *tert*-BUTYLOXYCARBONYL-L-PROLINE.

The Board of Editors thanks the contributors of the preparations and welcomes suggestions of changes that will improve
the usefulness of *Organic Syntheses*. The Editor-in-Chief urges
that the application of enzyme-catalyzed reactions, the continued use of new types of organometallic intermediates, and
the use of fermentation procedures for the preparation of organic
compounds be given serious consideration by his successors.
The attention of submitters of preparations is drawn to the
instructions on page vii. As initiated in Volume 50, *Organic
Syntheses* contains an insert listing unchecked preparations that
have been received during the preceding year. These are available from the Secretary's office for a nominal fee prior to checking. The Editorial Board of *Organic Syntheses*, with the help of
the Secretary, has issued a style guide for preparing submissions
to *Organic Syntheses* and the submitters are urged to follow its
suggestions in preparing their manuscripts. An innovation that
has been introduced in this volume is the inclusion of an Author
Index at the end of the volume as a replacement for the list of
contributors that has traditionally appeared at the beginning.

The Editor-in-Chief wishes to acknowledge the great assistance
of Dr. Albert I. Rachlin who, with his professional advice and

editing was instrumental in the successful completion of this volume. Additional thanks go to Miss Helen Rennie who typed much of the manuscript and checked proof; also to Dr. S. Kasparek who prepared the Contents and the Author and Subject Indexes.

Nutley, New Jersey   ARNOLD BROSSI
May 1973

CONTENTS

CONTENTS

3-ACETYL-2,4-DIMETHYLFURAN

(2,4-Dimethyl-3-furyl methyl ketone)

A. $(CH_3)_2S + BrCH_2C\equiv CH \longrightarrow (CH_3)_2\overset{+}{S}CH_2C\equiv CH\ Br^-$

B. $(CH_3)_2\overset{+}{S}CH_2C\equiv CH\ Br^-$
 $+$
 $CH_3COCH_2COCH_3 \xrightarrow[\text{ethanol, }\Delta]{C_2H_5ONa}$ [furan ring with CH_3, $COCH_3$, O, CH_3] $+ (CH_3)_2S$

Submitted by P. D. Howes and C. J. M. Stirling[1]
Checked by C. Reese, M. Uskoković, and A. Brossi

1. Procedure

Caution! These reactions should be performed in a hood because of the noxious odors.

A. *Dimethylprop-2-ynylsulfonium Bromide.* A mixture of 6.2 g. (0.1 mole) of dimethyl sulfide (Note 1), 11.9 g. (0.1 mole) of 3-bromopropyne (Note 2), and 10 ml. of acetonitrile (Note 3) is stirred magnetically for 20 hours (Note 4) in a darkened 100-ml. round-bottomed flask (Note 5) fitted with a calcium chloride drying tube. The resulting white, crystalline mass is filtered with suction and washed with three 50-ml. portions of dry ether (Note 6) to give 16.4 g. (90%) of the sulfonium salt, m.p. 105–106°. This material may be used in the next step without purification but, if desired, it may be recrystallized from ethanol-ether (Note 7) with minimal loss to give a product melting at 109–110°.

B. *3-Acetyl-2,4-dimethylfuran.* To a solution of 8.7 g. (0.087 mole) of acetylacetone (Note 8) in 175 ml. of $0.5M$ ethanolic sodium ethoxide (0.087 mole), contained in a 500-ml. round-bottomed flask fitted with a condenser topped with a calcium chloride drying tube, is added a solution of 15.75 g. (0.087 mole) of dimethylprop-2-ynylsulfonium bromide in

150 ml. of ethanol (Note 9). The mixture is refluxed until the odor of dimethyl sulfide is no longer appreciable (Note 10). The reaction flask is then fitted with a 30-cm. helix-packed column, and by heating the flask with a water bath, ethanol is distilled through the column (Note 11). The residue is treated with 200 ml. of ether, and the suspension is filtered. Ether is distilled from the filtrate at atmospheric pressure, and the residue is distilled to give 9.7 g. (81%) of 3-acetyl-2,4-dimethyl-furan (Notes 12 and 13), b.p. 90–95° (12 mm.), n^{24} D 1.4965.

2. Notes

1. Dimethyl sulfide was used as supplied by British Drug Houses.

2. 3-Bromopropyne, supplied by British Drug Houses, was distilled before use (b.p. 84–86°).

3. Acetonitrile (Matheson Coleman and Bell, spectral grade) was used without further treatment.

4. The maximum yield was obtained after 20 hours. Shorter reaction times give slightly lower yields.

5. If a brown glass flask is unavailable, an ordinary flask wrapped with aluminum foil may be used.

6. The ether was dried over sodium.

7. The salt was dissolved in 10 ml. of ethanol, 75 ml. of ether was added portionwise, and the mixture was allowed to stand overnight at room temperature.

8. Acetylacetone, supplied by British Drug Houses, was distilled before use (b.p. 137°).

9. The ethanol was dried with magnesium ethylate.[2]

10. About 6 hours is required on this scale.

11. Distillation through the packed column is essential to prevent loss of furan by codistillation with ethanol.

12. The product has i.r. absorption (neat) at 1690 cm.$^{-1}$ (ketone C=O) and n.m.r. peaks (CCl$_4$) at δ 2.20 (s, 3, COCH$_3$), 2.40 (s, 3, CH$_3$), 3.60 (s, 3, CH$_3$), and 7.40 (s, 1, furyl).

13. In a convenient modification of this procedure which gives the furan in 70–75% yield, the sulfonium salt is preformed in acetonitrile and, without isolation, the other reagents are added.

TABLE I
Furans Prepared Via Acetylenic Sulfonium Salts

$$R_3 \quad R_2 \quad \text{(furan ring)} \quad R_1$$

Sulfonium Salt	Addend	R_1	R_2	R_3	Yield, %[2]
$(CH_3)_2\overset{+}{S}CH_2C\equiv CH\ Br^-$	$CH_3COCH_2CO_2C_2H_5$	CH_3	$CO_2C_2H_5$	CH_3	86
$(CH_3)_2\overset{+}{S}CH_2C\equiv CH\ Br^-$	$CH_3COCH_2SO_2C_6H_4\text{-}p\text{-}CH_3$	CH_3	$SO_2C_6H_4\text{-}p\text{-}CH_3$	CH_3	78
$(CH_3)_2\overset{+}{S}CH_2C\equiv CH\ Br^-$	$C_6H_5COCH_2COC_6H_5$	C_6H_5	COC_6H_5	CH_3	72
$(CH_3)_2\overset{+}{S}CH_2C\equiv CC_6H_5\ Br^-$	$CH_3COCH_2CO_2C_2H_5$	CH_3	$CO_2C_2H_5$	$CH_2C_6H_5$	63
$(CH_3)_2\overset{+}{S}\underset{\underset{CH_3}{\mid}}{C}HC\equiv CH\ Br^-$	$CH_3COCH_2CO_2C_2H_5$	CH_3	$CO_2C_2H_5$	C_2H_5	50

3

3. Discussion

This procedure illustrates a recently published,[3] simple, general method for the synthesis of substituted furans. The scope of the reaction is shown in Table I. Many variations of this procedure are clearly possible.

The method described has some features in common with the well-known, but apparently little-used, Feist-Benary furan synthesis,[4] which uses an α-haloketone in place of the sulfonium salt. Acetylenic bromides suitable for preparing the sulfonium salts are readily available by well-documented procedures involving acetylenic organometallic compounds.

The mechanism of furan formation by this route is determined by the structure of the sulfonium salt; the course, hence the end product, is governed by whether an α-substituent is present. This must be considered when syntheses based on this procedure are being planned. Plausible mechanisms for the reaction have been suggested.[3]

Direct treatment of propargyl halides with β-dicarbonyl compounds and subsequent treatment of the products with zinc carbonate yields 2,3,5-trisubstituted furans.[5]

1. School of Physical and Molecular Sciences, University College of North Wales, Bangor, Caernarvonshire, U.K.
2. D. D. Perrin, W. L. F. Armarego, and D. R. Perrin, "Purification of Laboratory Chemicals," 1st ed., Pergamon Press, New York, 1966, p. 157.
3. J. W. Batty, P. D. Howes, and C. J. M. Stirling, *J. Chem. Soc.*, Perkin I, 65 (1973).
4. A. T. Blomquist and H. B. Stevenson, *J. Amer. Chem. Soc.*, **56**, 146 (1934).
5. K. E. Schulte, J. Reisch, and A. Mock, *Arch. Pharm*, **295**, 627 (1962).

2-ACETYL-6-METHOXYNAPHTHALENE

(6′-Methoxy-2′-acetonaphthone)

Submitted by L. Arsenijevic,[1] V. Arsenijevic,[1] A. Horeau,[2] and J. Jacques[2]
Checked by David Walba and Robert E. Ireland

1. Procedure

A 1-l. three-necked round-bottomed flask fitted with a mechanical stirrer is charged with 200 ml. of dry nitrobenzene (Note 1) followed by 43 g. (0.32 mole) of anhydrous aluminum chloride. The stirrer is started and after the aluminum chloride has dissolved, 39.5 g. (0.25 mole) of finely ground 2-methoxynaphthalene (nerolin) (Note 2) is added. One neck of the flask is fitted with a thermometer with the bulb in the solution, and the third neck of the flask is fitted with a 50-ml. pressure-equalizing addition funnel, carrying a drying tube that is attached to a gas trap. The flask is immersed in a slush of ice and water, and after the stirred solution has cooled to about 5°, 25 g. (22.6 ml., 0.32 mole) of redistilled acetyl chloride (Note 3) is added dropwise from the funnel in 15–20 minutes. The stirring and the addition rate are adjusted so that the temperature holds between 10.5 and 13° (Note 4). After addition of the acetyl chloride is complete, the flask is kept immersed in the ice water while stirring is continued for 2 hours. The mixture is then allowed to stand at room temperature for at least 12 hours.

The reaction mixture is cooled in an ice bath and poured, with manual stirring, into a 600-ml. beaker containing 200 g. of crushed ice, and then treated with 100 ml. of concentrated hydrochloric acid. The resulting two-phase mixture is transferred to a 1-l. separatory funnel, to which about 50 ml. of chloroform is also added (Note 5). The chloroform-nitrobenzene layer is separated and washed with three 100-ml. portions of water. The organic layer is then transferred to a 2-l. round-bottomed flask, and is steam-distilled. A fairly rapid flow of steam is used, and the distillation flask is heated in an oil bath at about 120°. After about 3 hours (3–4 l. of water) the distillation is stopped, and the residue in the flask is allowed to cool. Residual water in the flask is decanted from the solid organic material and extracted with chloroform. The solid residue in the flask is dissolved in 100 ml. of chloroform, separated from any water left in the flask, and the chloroform layers are combined and dried over anhydrous $MgSO_4$. The chloroform is stripped on a rotary evaporator and the solid residue, weighing 50–65 g. (still slightly wet with chloroform), is distilled under vacuum (Note 6). The receiving flask should be immersed in ice water, and the fraction boiling about 150–165° (0.02 mm.) is collected (Note 7).

The yellow distillate (*ca.* 40 g., m.p. 85–95°) is recrystallized from 75 ml. of methanol, cooled in an ice bath (Note 8) and filtered. The yield of white crystalline 2-acetyl-6-methoxynaphthalene (Note 9) is 22.5–24 g. (45–48%), m.p. 106.5–108° (lit. 104–105°).[3]

2. Notes

1. The nitrobenzene may be dried by distilling the first 10% and using the residue directly, or by standing over anhydrous calcium chloride overnight and filtering.

2. 2-Methoxynaphthalene (Matheson Coleman and Bell), m.p. 71.5–73°, was used without further purification.

3. Acetic anhydride can be used instead of acetyl chloride. However, it is then necessary to take two molecular equivalents of aluminum chloride per mole of anhydride and the amount of nitrobenzene must be increased by about 30%. About the same yield of ketone is obtained.

4. Temperature control is very important (see discussion).

5. The addition of chloroform is not always indispensable, but it is very useful to prevent emulsification and to facilitate

separation of the nitrobenzene layer. The reaction vessel and beaker are rinsed with this chloroform before it is added to the nitrobenzene layer. If an emulsion does form and phase separation becomes inconveniently slow, as much nitrobenzene as possible is withdrawn, and the emulsion and water layers are filtered by suction through a Celite cake wet with chloroform. The phases should then separate easily.

6. The material may be distilled from a distillation flask of the two bulb type described in *Org. Syn.*, Coll. Vol. **3**, 133 (1955), or from a small Claisen flask.

7. Care must be taken to prevent solidification and possible blocking in the condenser. A small burner may be used to keep the adapter just hot enough to melt the distillate.

8. If the methanol is cooled below 0°, the 1-acetyl isomer that is formed during the reaction, will also crystallize with the product.

9. N.m.r. (CDCl$_3$): δ 2.65 (s, 3, COCH_3, 3.92 (s, 3, OCH_3), 7.20 (m, 4, ArH), 7.80 (m, 1, ArH), 8.30 (m, 1, ArH).

3. Discussion

The procedure herein described is a modification of that of Haworth and Sheldrick,[3] the efficiency of which has been confirmed by many authors.[4-7] 2-Acetyl-6-methoxynaphthalene has also been prepared by the reaction of methylzinc iodide on 6-methoxy-2-naphthoyl chloride.[8]

In this reaction, nitrobenzene as solvent has an important function because it causes acylation to occur predominantly at the 6-position, whereas 1-acetyl-2-methoxynaphthalene is the principal product when carbon disulfide is used. The main feature of this procedure is the particular attention given to temperature control in order to obtain reliable results. It has been observed that the ratio of 6-acetylated to 1-acetylated nerolin is dependent on the temperature, the lower temperatures favoring 1-acetylation. Below 0° the yield of 6-acetylated product is only 3–10%. At higher temperatures the 6-acetylated product predominates, but an increased amount of tarry material is formed.

1. Department of Organic Chemistry, Faculty of Pharmacy, Belgrade, Yugoslavia.
2. Organic Chemistry of Hormones Laboratory, College of France, 75231 Paris 5, France.
3. R. D. Haworth and G. Sheldrick, *J. Chem. Soc. London*, 864 (1934).
4. R. Robinson and H. N. Rydon, *J. Chem. Soc. London*, 1394 (1939).
5. L. Novak and M. Protiva, *Collect. Czech. Chem. Commun.*, **22**, 1637 (1957).
6. N. P. Buu Hoi, D. Lavit, and J. Collard, *Croat. Chem. Acta*, **29**, 291 (1957) [*C. A.*, **53**, 2170b (1959)].
7. J. Cason and H. Rapoport, "Laboratory Text in Organic Chemistry," 2nd ed., Prentice-Hall, Engelwood Cliffs, N. J., 1962, p. 439.
8. K. Fries and K. Schimmelschmidt, *Ber.*, **58**, 2835 (1925).

ADAMANTANONE

(Tricyclo[3.3.1.1^{3,7}]decanone)

$98\% \ H_2SO_4 \quad | \quad 80°$

Submitted by H. W. GELUK and V. G. KEIZER[1]
Checked by L. FOLEY, W. JACKSON, and A. BROSSI

1. Procedure

Caution! This procedure should be carried out in an efficient hood to avoid exposure to sulfur dioxide.

A 1-l. three-necked round-bottomed flask equipped with an efficient mechanical stirrer (Note 1), a thermometer, and a vent

(Note 2) is placed in a water bath and charged with 600 ml. of 98% sulfuric acid (Note 3). Powdered adamantane,[2] 100 g. (0.735 mole) is added all at once to the stirred acid, and the mixture is then heated rapidly (by means of the water bath) to an internal temperature of 70°. The internal temperature is then raised gradually to 80° over a 2-hour period (Note 4) while vigorous stirring is maintained (Note 5). Stirring at 80° is continued for 2 hours longer, and then the temperature is raised to 82°. When almost all the adamantane is dissolved, the residual sublimed material should be scraped and rinsed from the walls of the flask (Notes 1 and 6). When g.l.c. analysis indicates that 2–3% of adamantanol is present (Note 6), the hot reaction mixture is poured immediately onto 800 g. of crushed ice to yield a 1500-ml. suspension containing crude adamantanone.

A 750-ml. portion of this suspension of crude adamantanone is transferred to a 2-l. round-bottomed flask, equipped for steam distillation (Notes 7 and 8), which is placed in a heating mantle. The contents of the distillation flask are heated to 70°; then the external heating is turned off (Note 9) and steam is introduced carefully through both inlet tubes (Note 10). The two layers of distillate are separated, and the aqueous layer is extracted with two 75-ml. portions of methylene chloride. This steam distillation procedure is then repeated with the second half of the suspension of crude adamantanone. The organic extracts are combined, washed with 100 ml. of aqueous, saturated sodium chloride, and dried over anhydrous sodium sulfate, and the solvent is evaporated under reduced pressure. The yield of adamantanone is 52–53 g. (47–48%) (Notes 11, 12, and 13).

2. Notes

1. The stirrer should be placed just under the surface of the sulfuric acid. The flask should be filled to at least three-quarters of its volume so the sublimed material can be rinsed from the walls by vigorous stirring.

2. Sulfur dioxide generated during the reaction can escape

through the third neck of the flask. This neck should be wide enough to prevent clogging by the subliming adamantane.

3. The required amount of 98% sulfuric acid is prepared by adding 53 ml. of fuming sulfuric acid (65% free sulfur trioxide) to 530 ml. of 96% sulfuric acid or by adding 120 ml. of fuming sulfuric acid (30–33% free sulfur trioxide) to 480 ml. of 96% sulfuric acid.

4. During this part of the reaction a vigorous evolution of sulfur dioxide takes place. Care should be taken to ensure that the internal temperature does not rise too fast to prevent an uncontrollable increase in the evolution of the sulfur dioxide.

5. An adamantane layer floating on the sulfuric acid, caused by ineffective stirring or too rapid evolution of sulfur dioxide, can lead to heavy foaming and subsequent losses. This layer, if formed, may be brought into contact with the reaction mixture by increasing the speed of the stirrer or by lowering the bath temperature to 65–70°.

6. At this stage of the preparation the progress of the reaction should be monitored carefully by g.l.c. For this purpose samples are taken periodically from the reaction mixture, poured onto ice, extracted with methylene chloride, washed with water, and subjected to g.l.c. The submitters used an F and M Model 700 gas chromatograph equipped with a 2 m. by 3 mm. glass column, filled with 80/100 mesh Chromosorb-W-Hp impregnated with 9.5% Apiezon-L and 0.5% Carbowax-20 M, at 120° with a flame ionization detector and nitrogen as carrier gas. To obtain a good yield of a fairly pure product, the reaction should be stopped when the amount of 1-adamantanol remaining is between 2% and 3%. Prolonged heating will give a further reduction in the 1-adamantanol content but it should be emphasized that concurrently the yield of the adamantanone will be diminishing rapidly.

7. The flask is fitted with two inlet tubes, one narrow and adjustable positioned above the surface of the adamantanone suspension, and the other reaching half-way between the bottom of the flask and the surface. Both tubes are connected to the steam supply. The flask is connected through a splash head and a short adapter to a 1-l. three-necked round-bottomed

flask equipped with two very efficient reflux condensers stoppered with a wad of cotton. The receiving flask is charged with 100 ml. of benzene, which, during the course of the steam distillation, begins to reflux and thus rinses the adamantanone from the condensers. *Added in proof.* Because of the toxicity of benzene the submitters advise replacement of this solvent with ethyl acetate at this point and in Note 8.

8. The checkers found that it is advisable to use superheated steam and also to distill each portion twice. Thus when the first receiving flask fills with distillate, it is replaced with a second 1-l. round-bottom flask charged with 100 ml. of benzene and distillation is resumed.

9. Prolonged external heating will stimulate the foaming. Insulation by the heating mantle is sufficient to prevent extensive steam condensation in the distilling flask.

10. The short inlet tube above the surface (Note 7) will break the foam and enable smooth removal of the adamantanone. When foaming is very heavy, steam is introduced only through this short inlet tube.

11. The checkers found that an additional 3 g. of adamantanone could be obtained by extracting the combined contents of the two distillation flasks with methylene chloride, removing the solvent under reduced pressure and steam-distilling the residue. The total yield would then be 55–56 g. (50–51%).

12. The product is 97–98% pure by g.l.c. (Note 6) and is satisfactory for most purposes. If desired the adamantanone may be purified by either column chromatography (alumina, activity grade IV; eluent : ether) or by treatment with fuming sulfuric acid (20% free sulfur trioxide). For example 8.0 g. of adamantanone is added portionwise to 40 ml. of ice-cold fuming sulfuric acid. Then the solution is heated to 40° and maintained at this temperature for one hour. After pouring the mixture onto ice, the adamantanone is recovered by extraction with methylene chloride.

13. The infrared spectrum (KBr) shows a strong band at 1717 with minor peaks or shoulders at 1670, 1690, 1725, and 1740 cm.$^{-1}$; n.m.r. (CDCl$_3$) δ 2.04 (broad s, 12, $2CH + 5CH_2$) and 2.55 (broad s, 2, $2CHC{=}O$).

3. Discussion

Adamantanone is a very versatile starting material for the preparation of adamantane derivatives, especially those substituted at secondary carbon atoms.

The preparative method presented is a slight modification of that reported by Geluk and co-workers.[3,4] These authors also give a detailed account of the several reactions of adamantane, 1-adamantanol, and 2-adamantanol that take place in sulfuric acid.[3,5] Adamantanone can be prepared essentially as herein described starting with 1-adamantanol instead of adamantane[3]—the yield is better (70%) and the reaction time is shorter, but adamantane is a more suitable starting material.

Adamantane can also be oxidized to adamantanone with ozone[6] and the oxime can be made directly from adamantane by photooximation.[7]

1. Philips-Duphar B. V. Research Laboratories, Weesp, The Netherlands.
2. P. v. R. Schleyer, M. M. Donaldson, R. D. Nicholas, and C. Cupas, *Org. Syn.*, **42**, 8 (1962).
3. H. W. Geluk and J. L. M. A. Schlatmann, *Tetrahedron*, **24**, 5361 (1968).
4. H. W. Geluk and V. G. Keizer, *Syn. Commun.*, **2**, 201 (1972). This preparation appears in this volume of *Org. Syn.* by courtesy of Marcel Dekker, Inc., New York.
5. H. W. Geluk and J. L. M. A. Schlatmann, *Tetrahedron*, **24**, 5369 (1968).
6. S. Landa and L. Vodička, Czech. Patent 119348 (1966) [*C. A.*, **67**, P21508v (1967)].
7. E. Müller and G. Fiedler, *Chem. Ber.*, **98**, 3493 (1965).

AZETIDINE

(Trimethyleneimine)

$$HO(CH_2)_3NH_2 + 2CH_2{=}CHCO_2C_2H_5 \longrightarrow$$

$$HO(CH_2)_3N(CH_2CH_2CO_2C_2H_5)_2$$

$$HO(CH_2)_3N(CH_2CH_2CO_2C_2H_5)_2 + SOCl_2 \longrightarrow$$

$$Cl(CH_2)_3N(CH_2CH_2CO_2C_2H_5)_2 + HCl + SO_2$$

$$Cl(CH_2)_3N(CH_2CH_2CO_2C_2H_5)_2 + Na_2CO_3 \longrightarrow$$

$$\square N(CH_2)_2CO_2C_2H_5 + CH_2{=}CHCO_2C_2H_5 + NaCl + NaHCO_3$$

$$\square N(CH_2)_2CO_2C_2H_5 + KOH \longrightarrow$$

$$\square NH + CH_2{=}CHCOOK + C_2H_5OH$$

Submitted by DONALD H. WADSWORTH[1]
Checked by F. THOENEN, E. VOGEL, R. HOBI, and A. ESCHENMOSER

1. Procedure

A. *1-(2-Carbethoxyethyl)azetidine.* A solution of 150 g. (2.0 moles) of 3-amino-1-propanol in 500 g. (5.0 moles) of ethyl acrylate (Note 1) is refluxed for 2 hours in a 1-l. round-bottomed flask. Subsequent vacuum stripping of the excess ethyl acrylate at steam temperature furnishes 548 g. (99%) of crude diethyl 3-N-(3-hydroxypropyl)iminodipropionate. A stirred, cooled solution of this material (548 g.) in 1 l. of chloroform and 10 ml. of dimethylformamide is treated dropwise with 262 g. (2.2 moles) of thionyl chloride. By cooling with an ice bath and controlling the addition rate, the reaction temperature is maintained below 40° (Note 2). After the addition is complete, the reaction mixture is stirred for 30 minutes at room temperature and poured slowly into a slurry of 340 g. of sodium bicarbonate in 1 l. of water (Note 3). The organic layer is separated (Note 4) and dried over sodium sulfate, and the solvent is removed under reduced pressure, below 50°, to furnish 570 g. (97%) of crude diethyl 3-N-(3-chloropropyl)iminodipropionate. A mixture of 100 g. of this crude material, 200 g. of anhydrous, powdered sodium carbonate (Note 5), and 10.0 g. of pentaerythritol (Note 6) in 200 ml. of diethyl phthalate is placed in a 500-ml. round-bottomed flask fitted with a vacuum-distillation head and an effective stirrer. The system is evacuated through a trap of sufficient capacity to contain 50 ml. of liquid, and the product is distilled by heating the stirred suspension with a heating mantle at 10–15 mm. By proper adjustment of the heat source, the distillation temperature is maintained below 150° to mini-mize codistillation of diethyl phthalate. The distillate is collected until the head temperature cannot be kept below 150°. Redistillation of the resulting crude product through a 4-in. Vigreux column furnishes 34.0 g. (57–68%) of 1-(2-carbethoxy-ethyl)azetidine, b.p. 86–87° (12 mm.), 99% pure by v.p.c. (Note 7).

B. *Azetidine.* A stirred mixture of 38 g. (0.68 mole) of potassium hydroxide pellets in 100 ml. of white mineral oil (Note 8) is heated to 140–150° in a four-necked 500-ml. round-bottomed flask, fitted with an air-driven Hershberg stirrer, a thermometer, a dropping funnel, and a 6-in. Vigreux column fitted with a vacuum-distillation head. The flask is removed from the heat source, and 50 g. (0.32 mole) of purified 1-(2-carbethoxyethyl)azetidine is added dropwise at a rate sufficient to maintain the reaction temperature at 150° (Note 9). After addition is complete, the reaction mixture is heated to 200° at 50 mm. to remove all traces of ethanol (Note 10). The flask is fitted with a distillation head and a nitrogen bubbler, and the distillation is resumed at atmospheric pressure until azetidine distills (210° maximum pot temperature) (Note 11). The result-ing product (19.6 g., 85% purity) is dried over potassium hydroxide and redistilled through a short Vigreux column to furnish 14.5–15.8 g. (80–87%) of purified azetidine, b.p. 62–63° (Note 12).

2. Notes

1. Eastman Organic Chemicals, practical grade, 3-amino-1-propanol, and ethyl acrylate were used. The checkers used 3-amino-1-propanol, FLUKA purum.

2. If the reaction temperature is not controlled, a tarry by-product is formed. By dissolving the crude aminochloride in petroleum ether, the impurity is separated as an insoluble tar.

3. Considerable foaming occurs during neutralization, which is best accomplished in a 4-l. beaker with rapid stirring.

4. Any excess of insoluble salt should be filtered from the reaction mixture before the chloroform layer is separated. The checkers observed no insoluble salt at this point.

5. Baker and Adamson reagent grade sodium carbonate powder was used. The checkers found that the use of crystalline anhydrous sodium carbonate lowered the yield.

6. The presence of some hydroxylic material appears to be necessary to ensure reproducibility of this step, and the submitters have found empirically that pentaerythritol is very effective for this purpose.

7. The crude distillate can be used directly without intermediate isolation of 1-(2-carbethoxyethyl)azetidine to furnish a 50% yield of azetidine, assuming that all of the crude material was 1-(2-carbethoxyethyl)azetidine and running the reaction as described for the pure material.

8. Fisher Scientific Company Paraffin Oil, N.F., Saybolt Viscosity 125/135 and Mobil Oil Corp. S/V Industrial White Oil Number 320, Saybolt Viscosity 200/210 gave comparable results.

9. If the reaction does not start immediately, 5 ml. of ethanol may be added as an initiator.

10. Any ethanol that is not removed will contaminate the product. Although careful distillation will separate azetidine and ethanol, a considerable yield loss is encountered.

11. The submitters used a higher pot temperature (230–275°). The checkers, however, recommend the lower temperature to minimize losses of azetidine.

12. N.m.r. (CDCl$_3$) δ 1.85 (s, 1, NH), 2.0–2.6 (m, 2, CH_2), 3.68 (t, 4, $J = 8$ Hz., 2 CH_2).

3. Discussion

Azetidine has been prepared by the following procedures: cyclization of 3-bromopropylamine with potassium hydroxide (low yield);[2] cleavage of 1-*p*-toluenesulfonylazetidine with sodium and refluxing *n*-amyl alcohol (85–100% yield)[3,4] or sodium and liquid ammonia (30% yield);[5] hydrogenolysis of 1-benzylazetidine (50%);[6] cyclization of diethyl 3-N-(3-chloropropyl)iminodipropionate with sodium carbonate without solvent (60–70% yield).[7] A review of methods for preparing azetidine has been published.[8]

All other preparations of azetidine suffer either from low yields, arduous preparative procedures, or cumbersome purification operations. In this procedure, cyclization is accomplished in relatively concentrated solutions, and azetidine is obtained directly from the cleavage reaction in a state sufficiently pure for most applications. In addition, stable 1-(2-carbethoxyethyl)-azetidine can be prepared in advance, and the air-sensitive azetidine can be formed readily as needed by a one-step procedure.

1. Eastman Kodak Company, Rochester, New York 14650.
2. S. Gabriel and J. Weiner, *Ber.*, **21**, 2669 (1888).
3. C. C. Howard and W. Markwald, *Ber.*, **32**, 2031 (1899).
4. W. F. Vaughn, R. S. Klonowski, R. S. McElhinney, and B. B. Millward, *J. Org. Chem.*, **26**, 138 (1961). Many investigators have had difficulty in repeating this procedure. Apparently the azetidine tends to be swept out of the reaction mixture with the hydrogen.
5. A. B. Burg and C. D. Good, *J. Inorg. Nucl. Chem.*, **2**, 237 (1956).
6. R. S. Klonowski, Ph.D. Dissertation, University of Michigan, 1959.
7. D. H. Wadsworth, *J. Org. Chem.*, **32**, 1184 (1967). Although this procedure will furnish good yields on a small scale, scale-up causes mechanical difficulties and lower yields.
8. James A. Moore, in "The Chemistry of Heterocyclic Compounds," Vol. 19, Part II, A. Weissberger (Ed.), Interscience Publishers, Inc., New York, 1964, Chap. VII.

BASE-INDUCED REARRANGEMENT OF EPOXIDES TO ALLYLIC ALCOHOLS: *trans*-PINOCARVEOL

[2(10)-Pinen-3α-ol]

Submitted by J. K. CRANDALL and L. C. CRAWLEY[1]
Checked by SHOICHIRO UYEO and WATARU NAGATA

1. Procedure

A dry, 300-ml. three-necked round-bottomed flask is fitted with an effective reflux condenser, a 50-ml. pressure-equalizing dropping funnel, a rubber septum, a magnetic stirring bar, and a nitrogen inlet tube on the top of the condenser to maintain a static nitrogen atmosphere in the reaction vessel throughout the reaction. The flask is flushed with dry nitrogen and then charged with 2.40 g. (0.034 mole) of diethylamine (Note 1) and 100 ml. of anhydrous ether (Note 2). The flask is immersed in an ice bath, the stirrer is started, and 25 ml. (0.035 mole) of 1.4M n-butyllithium in hexane (Notes 3 and 4) is added carefully through the rubber septum by means of a syringe. After stirring for 10 minutes, the ice bath is removed and 5.00 g (0.033 mole) of α-pinene oxide (2,3-epoxy pinane) (Note 5) in 20 ml. of anhydrous ether is added dropwise over a 10-minute period. The resulting mixture is heated to reflux with stirring for 6 hours (Note 11). After the clear homogeneous mixture is cooled in an ice bath, it is stirred vigorously while 100 ml. of water is added. The ether phase is separated and washed successively with 100 ml. portions of N hydrochloric acid, water, saturated aqueous sodium bicarbonate, and water. The aqueous phase and each washing are extracted twice with 50 ml. portions of ether, and the ethereal extracts are combined and dried over anhydrous magnesium sulfate. Evaporation of the solvent on a rotary evaporator yields a light-yellow, oily residue which is distilled through a short-path distillation head to give 4.50–4.75 g. (90–95%) of *trans*-pinocarveol as a colorless oil, b.p. 92–93° (8 mm.) n^{25}D 1.4955 (Notes 12 and 13).

2. Notes

1. Commercial diethylamine, b.p. 55–58°, purchased from Fisher Scientific Company, was distilled from calcium hydride before use. The checkers used material purchased from Kanto Chemical Company, Inc. (Japan) and distilled from sodium hydride.

2. The checkers used anhydrous ether distilled from sodium hydride before use.

3. The *n*-butyllithium in hexane solution was purchased from Foote Mineral Company. The checkers obtained their material from Wako Pure Chemical Industries Ltd. (Japan) and titrated it with $0.80M$ 2-butanol in xylene using 1,10-phenanthroline as indicator.[2] Care should be exercised in handling *n*-butyllithium solutions.

4. The submitters used about three molar equivalents of lithium diethylamide in about twice as much solvent. The checkers found that an amount of base slightly in excess of one molar equivalent was sufficient to convert the epoxide to exocyclic methylene alcohol of superior purity.

5. The submitters purchased α-pinene oxide from F.M.C. Corporation. However, since the compound is no longer available, the checkers prepared it from α-pinene as follows. Into a three-necked round-bottomed flask fitted with a 50-ml. dropping-funnel, a thermometer, and a magnetic stirring bar are placed 22.0 g. (0.102 mole) of *m*-chloroperbenzoic acid (Note 6), 11.0 g. (0.131 mole) of sodium bicarbonate, and 250 ml. of dichloromethane. The suspension is stirred with a powerful stirrer while being cooled with an ice-salt bath. To this mixture is added dropwise a solution of 13.6 g. (0.10 mole) of α-pinene (Note 7) in 20 ml. of dichloromethane at a rate such that the inner temperature is kept between 5–10° (Note 8). During

the addition, sodium *m*-chlorobenzoate begins to crystallize indicating that the reaction is proceeding. After completion of the addition, stirring is continued for 1 hour longer at the same temperature (Note 9). A solution of 5 g. of sodium sulfite in 50 ml. of water is added, and the mixture is stirred vigorously at room temperature for 30 minutes. Water (50 ml.) is added, and the dichloromethane phase is separated and washed with 100 ml. of 5% aqueous sodium carbonate. The two aqueous washings are extracted with 50 ml. of dichloromethane, and the organic solutions are combined and dried over anhydrous magnesium sulfate. Evaporation of the solvent on a rotary evaporator gives an oily residue that is distilled through a vacuum-jacketed column to yield 12.5–12.8 g. (82–85%) of α-pinene oxide as a colorless oil, b.p. 89–90° (28 mm.) (Note 10).

6. *m*-Chloroperbenzoic acid was obtained from F.M.C. Corporation. It was shown to be of 80% purity by titration.

7. Technical grade α-pinene, purchased from Wako Pure Chemical Industries Ltd. (Japan), was used without purification.

8. A more efficient cooling system, such as a dry ice-acetone bath, is necessary to shorten the addition time in large scale preparations.

9. Completion of the reaction may be checked by gas chromatography.

10. Gas chromatography of this product using a 1-m. column containing 5% KF-54 on Chromosorb W at 100° gave a single peak. The material gave the following n.m.r. spectrum (deuteriochloroform solution, internal tetramethylsilane reference): δ 0.95, 1.30, 1.33 (3s, 9, 3CH_3), 1.53–2.20 (m, 6, CH_2 and CH), 3.03 (m, 1, CH—O—C). The boiling point is reported to be 70–71° (12 mm.).[3]

11. Completion of the reaction may be checked by gas chromatographic analysis. Refluxing for prolonged periods can give saturated ketone as an impurity if excess base is used.

12. The reported[4] value is n^{20}D 1.4993.

13. The spectral properties are: i.r. (neat cm.$^{-1}$ 3360ms (OH), 1644vw (C=C), 893ms (C=CH$_2$); n.m.r. (deuteriochloroform solution, internal tetramethylsilane reference) δ 0.65, 1.28 (2s, 6, 2CH_3), 1.63–2.55 (m, 6, CH_2 and CH), 4.42 (d, 1, $J =$

7Hz., CH—OH), 4.82, 5.00 (2m, 2, C=CH_2). Purity of the product is greater than 98% as determined by gas chromatography using a Carbowax 20 M on 60-80 Chromosorb W column or a 1-m. column containing 5% KF-54 on Chromosorb W at 100°.

3.   Discussion

Pinocarveol has been prepared by the autoxidation of α-pinene,[5] by the oxidation of β-pinene with lead tetraacetate,[6] and by isomerization of α-pinene oxide with diisobutylaluminum,[7] lithium aluminum hydride,[8] activated alumina,[9] potassium *tert*-butoxide in dimethylsulfoxide,[10] and lithium diethylamide.[11] The present method is preferred for the preparation of pinocarveol, since the others give mixtures of products. It also illustrates a general method for converting 1-methylcycloalkene oxides into the corresponding exocyclic methylene alcohols.[11] The reaction is easy to perform, and the yields are generally high.

In general, the strong base isomerization of epoxides to allylic alcohols constitutes a useful synthetic reaction. Since the rearrangement is a highly specific process, it should be of value in organic synthesis. For example, there is a very high propensity for Hofmann-type eliminations to yield the least substituted double bond from unsymmetrically substituted epoxides.[12] There is also a large conformational effect arising from the operation of a *syn*-elimination mechanism which leads to specificity in eliminations of cyclic epoxides.

1. Department of Chemistry, Indiana University, Bloomington, Indiana 47401.
2. R. A. Ellison, R. Griffin, and F. N. Kotsonis, *J. Organometal. Chem.*, **36**, 209 (1972).
3. J. J. Ritter and K. L. Russell, *J. Amer. Chem. Soc.*, **58**, 291 (1936).
4. H. Schmidt, *Ber.*, **63**, 1129 (1930).
5. R. N. Moore, C. Golumbic, and G. S. Fisher, *J. Amer. Chem. Soc.*, **78**, 1173 (1956).
6. T. Sato, *Nippon Kagaku Zasshi*, **86**, 252 (1965).
7. P. Teisseire, A. Galfre, M. Plattier, and B. Corbier, *Recherches*, **15**, 52 (1966). *Chim. Fr.*, 2546 (1963).
8. Y. Chrétien-Bessière, H. Desalbres, and J. P. Monthéard, *Bull. Soc. Chim. Fr.*, 2546 (1963).

9. V. S. Joshi, N. P. Damodaran, and S. Dev, *Tetrahedron*, **24**, 5817 (1968).

10. J. P. Monthéard and Y. Chrétien-Bessière, *Bull. Soc. Chim. Fr.*, 336 (1968).

11. J. K. Crandall and L. H. Chang, *J. Org. Chem.*, **32**, 435 (1967); J. K. Crandall and L. H. C. Lin, *J. Org. Chem.*, **33**, 2375 (1968).

12. R. P. Thummel and B. Rickborn, *J. Org. Chem.*, **36**, 1365 (1971); R. P. Thummel and B. Rickborn, *J. Amer. Chem. Soc.*, **92**, 2064 (1970); B. Rickborn and R. P. Thummel, *J. Org. Chem.*, **34**, 3583 (1969).

2-*tert*-BUTYL-1,3-DIAMINOPROPANE

(1,3-Propanediamine, 2-*tert*-butyl)

$$CH_2(CN)_2 + (CH_3)_3CCl \xrightarrow[\text{nitromethane}]{AlCl_3} (CH_3)_3CCH(CN)_2$$

$$\xrightarrow[\text{tetrahydrofuran}]{B_2H_6} \xrightarrow{HCl} (CH_3)_3CCH(CH_2\overset{+}{N}H_3\overset{-}{Cl})_2$$

$$\xrightarrow{NaOH} (CH_3)_3CCH(CH_2NH_2)_2$$

Submitted by R. O. HUTCHINS[1] and B. E. MARYANOFF
Checked by ABRAHAM PINTER and RONALD BRESLOW

1. Procedure

A. *tert-Butylmalononitrile.* In a dry 2-l. three-necked flask equipped with a thermometer, mechanical stirrer, and a Claisen adapter fitted with a dropping funnel and condenser protected with a drying tube is placed 200 ml. of nitromethane (Note 1). The flask is cooled to 0° in an ice-salt bath (Note 2), and anhydrous powdered aluminum chloride (90.0 g., 0.68 mole) is added with slow stirring through a powder funnel that temporarily replaces the thermometer. The temperature may rise to *ca.* 50° but quickly drops to 0°. Malononitrile (45.0 g., 0.68 mole) (Note 3) in 50 ml. of nitromethane is added through the dropping funnel at a rate such that the temperature is kept below 10° (approximately one hour), followed by slow dropwise addition of *tert*-butyl chloride (Note 4) (150 g., 1.62 moles) in 50 ml. of nitromethane at a rate such that the temperature is maintained below 10° (approximately 3–4 hours). The reaction mixture is stirred at 0–5° for 10 hours, and then 1 l. of saturated sodium bicarbonate (*ca.* 80 g. in 1000 ml. of water) is added slowly and cautiously (still in the cold) while the temperature is kept below 10°. The mixture is then poured into a 3 or 4 l. beaker and solid sodium bicarbonate (*ca.* 100 g.) is added with stirring in *small* portions. The organic phase is separated and the aqueous layer extracted with three half-volume portions of methylene chloride. The methylene chloride extracts are concentrated at reduced pressure on a rotary evaporator, combined with the organic phase, and concentrated further, and the resultant brown oil is subjected to steam distillation. The first fraction is collected using a cold water condenser until solidification is observed in the condenser, at which time warm water is passed through the condenser and the receiver is changed (Note 5). The product is collected until occasional passage of cold water through the condenser no longer causes apparent solidification. At this point, the receiver is changed again, and a third fraction totaling about 500 ml. of distillate is obtained. The middle fraction is cooled in ice and filtered with vacuum through a medium fritted glass funnel to afford 48–50 g. of light-yellow product. Extraction of the filtrate with two half-volumes of ether followed by evaporation gives an additional small amount (*ca.* 2 g.) of product. A further small crop of product may be gleaned from the first and third steam distillation fractions by separating any organic phase, followed by removing distillable material on a rotary evaporator and cooling and filtering the resulting solid. The total combined yield of crude product is 54–58 g. (65–70%). Further purification is accomplished by careful sublimation at 80–90 mm. (*ca.* 85°) to give 52–56.5 g. (63–68%) of white, waxy dinitrile, m.p. 76–79° with softening at 71°.

B. *2-tert-Butyl-1,3-diaminopropane.* A dry 500-ml. three-necked flask fitted with a magnetic stirrer, nitrogen inlet, dropping funnel, and condenser attached to an acetone gas trap (Note 6) is flushed with dry nitrogen for 30 minutes and charged with purified *tert*-butylmalononitrile (30.5 g., 0.25 mole), sodium borohydride (17.5 g., 98% assay, 0.453 mole), and 150 ml. of dry tetrahydrofuran (Note 1). A dry nitrogen atmosphere is maintained while boron trifluoride etherate (85.3 g., 0.60 mole) (Note 7) in 50 ml. of dry tetrahydrofuran is added dropwise, with slow magnetic stirring at a rate that permits gentle reflux. The addition takes about 4 hours (Note 8). The mixture is stirred for an additional 90 minutes, hydrolyzed by the very cautious dropwise addition of 30 ml. of concentrated hydrochloric acid, and transferred to a 1-l. flask with rinsing by 100 ml. of tetrahydrofuran. The solution is evaporated to dryness on a rotary evaporator to yield a dry, white, solid mass to which is added a small portion (*ca.* 10 ml.) of 125 ml. of aqueous 40% sodium hydroxide (*w/w*). The mixture is triturated with a glass rod and warmed on a steam bath until a reaction begins. Heat is generated, and white smoke is evolved. The reaction is controlled by cooling in an ice bath. When the reaction appears to have subsided at room temperature, the trituration is repeated cautiously until all 125 ml. of solution is added. The resulting mixture is warmed for 30 minutes on a steam bath with occasional stirring, cooled, and filtered with vacuum. The solid material is washed with ten 20-ml. portions of ether. The combined filtrate is separated, the aqueous phase extracted with three 100-ml. portions of ether, and the combined ethereal extracts and organic phase dried over anhydrous sodium sulfate. The drying agent is filtered and washed with ether, and the filtrate concentrated on a rotary evaporator. The residue is fractionally distilled at reduced pressure through a short Vigreux column. After removal of residual solvent and collection of a small forerun, 11.5–15.5 g. (36–48%) of product is obtained, b.p. 92.5–95° (21 mm.), 96–98° (27 mm.), n^{28}D 1.4570–1.4585. Analysis of the product by gas-liquid chromatography (OV-1 column) indicated the purity to be *ca.* 95% (Note 9).

2. Notes

1. Fisher Scientific Company certified reagent grade was used without further purification from a freshly opened bottle.

2. As a convenience, the reaction may be conducted in a

refrigerated room in which case it will be unnecessary to replenish the ice during the course of the reaction.

3. Malononitrile was obtained from Eastman Organic Chemicals and distilled prior to use, b.p. 80–82° (3 mm.).

4. The commercial material was distilled prior to use, b.p. 50–51°.

5. Passing warm water through the condenser prevents blockage by solid product; otherwise pressure may build and force the joints apart. As an alternative, steam may be passed through the condenser periodically.

6. Impure diborane is a hazardous material and may combust explosively on contact with air. Therefore, precautions must be taken to prevent escape from the reaction.

7. The commercial material was distilled prior to use, b.p. 59–60° (20 mm.). The material employed should not be more than a few days old.

8. The duration of the reaction is important and should not be curtailed. For example, if one is operating on a one-tenth scale, the reaction should be heated at gentle reflux for an additional 3.5 hours after the addition (25 minutes).

9. A sample of the compound, collected by g.l.p.c., analyzed correctly for carbon, hydrogen, and nitrogen. Two solid derivatives were prepared in good yield and both analyzed correctly.

3. Discussion

This preparation illustrates the alkylation of malononitrile under acid-catalyzed conditions, and the use of diborane for the reduction of a dinitrile to a diamine. The procedure for the preparation of *tert*-butylmalononitrile has been outlined briefly by Boldt and co-workers.[2] The generation of diborane *in situ* and the general method for nitrile reduction is that described by Brown and co-workers.[3] Attempts to reduce the dinitrile to the diamine by other methods including catalytic hydrogenation (5% rhodium on alumina, 5 atm.), lithium aluminum hydride, and lithium aluminum hydride-aluminum chloride were singularly unsuccessful.

1. Department of Chemistry, Drexel University, Philadelphia, Pennsylvania, 19104.
2. P. Boldt and W. Thielecke, *Angew. Chem. Int. Ed. Engl.*, **5**, 1044 (1966); P. Boldt and L. Schulz, German Patent 1,200,298 (1965) [*C.A.* **63**, 16221e (1965)]; P. Boldt and L. Schulz, *Naturwissenschaften*, **51**, 288 (1964).
3. H. C. Brown and B. C Subba Rao, *J. Amer. Chem. Soc.*, **82**, 681 (1960); and more recently, H. C. Brown, P. Heim, and N. M. Yoon, *J. Amer. Chem. Soc.*, **92**, 1637 (1970).

tert-BUTYLOXYCARBONYL-L-PROLINE

(1,2-Pyrrolidinedicarboxylic acid, 1-*tert*-butyl ester, L-)

$$\text{proline} \xrightarrow[\substack{(CH_3)_2NC(=NH)N(CH_3)_2 \\ \text{dimethyl sulfoxide}}]{(CH_3)_3COCO_2C_6H_5} \text{Boc-proline} + C_6H_5OH$$

Submitted by ULF RAGNARSSON, SUNE M. KARLSSON, BENGT E. SANDBERG, and LARS-ERIC LARSSON[1]
Checked by S. WANG and A. BROSSI

1. Procedure

A 1-l. Erlenmeyer flask (Note 1) equipped with a magnetic stirrer, and a thermometer is charged with 115 g. (1.0 mole) of L-proline (Note 2) and 500 ml. of dimethyl sulfoxide (Note 3). To the stirred suspension are added simultaneously, over 5 minutes, 115 g. (1.0 mole) of 1,1,3,3-tetramethylguanidine (Note 4) and 214 g. (1.10 moles) of *tert*-butyl phenyl carbonate (Note 5). The proline dissolves completely within a few minutes in an exothermic reaction, the temperature of which reaches a maximum of 50–52° after 10–15 minutes. After stirring for 3 hours, the clear reaction mixture is transferred to a 6-l. separatory funnel and shaken with 2.2 l. of water and 1.8 l.

of ether (Note 6). The aqueous layer, after being washed with 500 ml. of ether, is acidified to pH 3.0 by the addition of 10% sulfuric acid (Note 7), which generally causes partial crystallization of the product. The acidified solution, including the solid, is extracted with three 600-ml. portions of a mixture of equal volumes of ethyl acetate and ether. The combined extracts are washed with three 25-ml. portions of water, dried over magnesium sulfate, filtered, and evaporated with a rotary evaporator at a bath temperature not exceeding 40°. After drying in a vacuum oven at 50°, the crude *tert*-butyloxycarbonyl-L-proline weighs 202 g., m.p. 129–132°. It is recrystallized by solution in 300 ml. of hot ethyl acetate, and clarified by filtration and the addition of 1 l. of petroleum ether (40–60°). The product, after drying under vacuum at 50°, weighs 179–193 g. (83–90%), m.p. 132–134°, $[\alpha]^{25}_D$ −59.84° to −61.6° ($c = 1$, glacial acetic acid) (Notes 8, 9, and 10).

2.   Notes

1. A three-necked flask equipped with a U-tube may also be used for the reaction.

2. The submitters used L-proline obtained from Tanabe Seiyaku Company, Ltd., Osaka, Japan and checked its purity by the method of Manning and Moore.[2] The L-proline used by the checkers was obtained from Ajinomoto Company, New York.

3. Dimethyl sulfoxide, Fisher Scientific Company, was used without further purification.

4. 1,1,3,3-Tetramethylguanidine, b.p. 159–160°, was used without further purification. The submitters obtained their material from Schuchardt, Munich, Germany and the checkers obtained theirs from Pfaltz and Bauer, New York.

5. *tert*-Butyl phenyl carbonate furnished by Ega-Chemie KG, Steinheim, Germany, was used by the submitters. The checkers used material obtained from Aldrich Chemical Company, Inc.

6. The pH of the aqueous layer was 7.2. If the pH, as measured with a pH meter, is not between 7 and 8, it should be

adjusted to within these limits by the addition of either 10% sulfuric acid or 1,1,3,3-tetramethylguanidine. The submitters worked up the reaction by the following alternate but less convenient method. The reaction mixture was poured into 1.25 l. of M sodium bicarbonate solution, 500 ml. of ether, and sufficient water (*ca.* 1 l.) to give two clear phases. The pH, which was 8.9, was adjusted to 8.0 by the addition of solid potassium bisulfate with stirring.

7. 10% Sulfuric acid ($1.1M$) was prepared by diluting 25 ml. of concentrated sulfuric acid with 398 ml. of water. The submitters used solid potassium bisulfate for acidification to pH 3.0.

8. An additional 5 g. of product, m.p. 127–129°, may be obtained from the mother liquor.

9. Some reported yields, melting points, and rotations are: 55%, 136–137°, and $[\alpha]^{25}$D —60.2° (2.011 in acetic acid)[3]; 96%, 134–136°, $[\alpha]^{18-25}$ 578 —68.5° ($c = 1$, acetic acid)[4]; 95%, 132–134°, $[\alpha]$ 578 —62.5° (acetic acid)[5]; 90%, 133–135°, $[\alpha]^{25}$D —60.4° ($c = 2.2$ in acetic acid)[6]; 93%, 134–135°, no rotation reported.[7]

10. N.m.r. (DMSO, internal TMS): δ 1.38 (s, 9, $3CH_3$), 1.92 (m, 4, $2CH_2$), 3.31 (t, 2, CH_2N), 4.05 (t, 1, CHN), 12.3 (s, 1, CO_2H). Analysis calculated for $C_{10}H_{17}NO_4$: C, 55.80; H, 7.96; N, 6.51. Found: C, 55.81; H, 7.95; N, 6.44.

3. Discussion

Since their introduction by McKay and Albertson[8] and Anderson and McGregor,[3] *tert*-butyloxycarbonyl amino acids have been prepared by several different methods. The simplest procedure[6] requires working with large quantities of phosgene. Another very good method,[5] but one that has not found wide application, involves the use of *tert*-butyloxycarbonylfluoride that is not commercially available. Up to the present time the most useful reagent has been *tert*-butyloxycarbonylazide for which good procedures[9,10] are available. The excellent method of Schnabel[4] and the one more recently reported[7] are based on

TABLE I

OTHER Boc[a] AMINO ACIDS SYNTHESIZED BY THIS PROCEDURE

Compound	Solvent	Temperature, °	Time, Hr.	Yield, %	Remarks
Boc-Ala[b]	DMSO[c]	25	40	58	
Boc-Ala[b]	DMSO[c]	40	40	79	
Boc-Asn[d]	DMSO[c]	25	66	70	2 equiv. TMG[e]
Boc-Asp[f]	DMSO[c]	25	18	89	2 equiv. TMG
Boc-Cys (Bzl)[g]	DMSO[c]	25	40	62	CHA salt[h]
Boc-Cys (Bzl)[g]	DMSO[c]	40	40	78	CHA salt
Boc-Gln[i]	DMSO[c]	50	48	62	2 equiv. TMG, DCHA salt,[j] continuous extraction with ethyl acetate
Boc-Glu[k]	DMSO[c]	25	2.5	80	2 equiv. TMG
Boc-Ile[l]	DMSO[c]	40	72	72	
Boc-Leu[m]	DMSO[c]	25	48	73	Calc. as hemihydrate
Boc-Met[n]	DMSO[c]	25	24	86	81% solid +5% DCHA salt
Boc-Phe[o]	DMSO[c]	25	40	81	DCHA salt
Boc-Phe[o]	DMSO[c]	25	96	88	DCHA salt
Boc-Phe[o]	DMSO[c]	40	18	81	DCHA salt
Boc-Phe[o]	DMF[p]	25	48	59	DCHA salt
Boc-Phe[o]	D−W[q]	25	48	5	DCHA salt
Boc-Pro[r]	DMSO[c]	25	2.5	90	
Boc-Pro[r]	DMF[p]	25	2.5	92	
Boc-Pro[r]	D−W[q]	25	21	84	
Boc-Thr[s]	DMSO[c]	25	67	66	DCHA salt
Boc-Val[t]	DMSO[c]	25	71	77	2 equiv. TMG, 65% solid +12% DCHA salt

[a] tert-Butyloxycarbonyl.   [b] L-Alanine.   [c] Dimethylsulfoxide.   [d] L-Asparagine.   [e] 1,1,3,3-Tetramethylguanidine.   [f] L-Aspartic acid.   [g] S-Benzyl-L-cysteine.   [h] Cyclohexylamine.   [i] L-Glutamine.   [j] Dicyclohexylamine.   [k] L-Glutamic acid.   [l] L-Isoleucine.   [m] L-Leucine.   [n] L-Methionine.   [o] L-Phenylalanine.   [p] Dimethylformamide.   [q] Dioxane-water 1:1.   [r] L-Proline.   [s] L-Threonine.   [t] L-Valine.

this reagent. Of the procedures for the preparation of *tert*-butyloxycarbonylazide one,[9] which is readily adaptable for large-scale operations, involves three steps and the other,[10] a two-step process, is more suitable for small-scale work.

Our procedure[11] represents a simplication in that *tert*-butyl phenyl carbonate, which is used as a starting material, is the first intermediate in the three-step synthesis[9] of *tert*-butyloxycarbonylazide. This reagent is easy to prepare in quantity in the average laboratory and it is also commercially available in bulk (Note 5). Further, 1,1,3,3-tetramethylguanidine is inexpensive and the experimental operations are extremely simple.

Proline dissolves readily in dimethyl sulfoxide. Some other amino acids which are less-soluble require longer reaction times and, in some instances, other solvents.[11] These details and the scope of the reaction are illustrated in Table I.

1. Biokemiska Institutionen, Uppsala Universitet, Box 531, S-751 21 Uppsala 1, Sweden.
2. J. M. Manning and S. Moore, *J. Biol. Chem.*, **243**, 5591 (1968).
3. G. W. Anderson and A. C. McGregor, *J. Amer. Chem. Soc.*, **79**, 6180(1957).
4. E. Schnabel, *Justus Liebigs Ann. Chem.*, **702**, 188 (1967).
5. E. Schnabel, H. Herzog, P. Hoffmann, E. Klauke, and I. Ugi, "Peptides 1968," E. Bricas (Ed.), North-Holland Publishing Company, Amsterdam, 1968, p. 91.
6. S. Sakakibara, I. Honda, K. Takada, M. Miyoshi, T. Ohnishi, and K. Okumura, *Bull. Chem. Soc. Jap.*, **42**, 809 (1969).
7. A. Ali, F. Fahrenholz, and B. Weinstein, *Angew. Chem.*, **84**, 259 (1972). [*Angew. Chem. Int. Ed. Engl.*, **11**, 289 (1972)].
8. F. C. McKay and N. F. Albertson, *J. Amer. Chem. Soc.*, **79**, 4686 (1957).
9. L. A. Carpino, B. A. Carpino, P. J. Crowley, C. A. Giza, and P. H. Terry, *Org. Syn.*, **44**, 15 (1964).
10. M. A. Insalaco and D. S. Tarbell, *Org. Syn.*, **50**, 9 (1970).
11. U. Ragnarsson, S. M. Karlsson, and B. E. Sandberg, *Acta Chem. Scand.*, **26**, 2550 (1972).

DIAMANTANE: PENTACYCLO[7.3.1.1^{4,12}.0^{2,7}.0^{6,11}]TETRADECANE

(3,5,1,7-[1,2,3,4]Butanetetraylnaphthalene, decahydro)

$$\xrightarrow[\text{BF}_3 \cdot \text{O(C}_2\text{H}_5)_2]{\text{CoBr}_2 \cdot 2\text{(C}_6\text{H}_5)_3\text{P}}$$

$$\xrightarrow[\text{CH}_3\text{CO}_2\text{H, HCl}]{\text{H}_2,\ \text{PtO}_2}$$

$C_{14}H_{20}$

Tetrahydro-Binor-S

$C_{14}H_{20}$
Tetrahydro-Binor-S

$$\xrightarrow[\text{CS}_2\ \text{or}\ \text{C}_6\text{H}_{12}]{\text{AlBr}_3}$$

Submitted by Tamara M. Gund, Wilfried Thielecke,
and Paul v. R. Schleyer[1]
Checked by H. Gurien, R. Regenye, and A. Brossi

1. Procedure

A. *Binor-S*.[2] A 2-l. 3-necked flask equipped with Teflon
sleeves (Note 1), a thermometer, a condenser, a dropping funnel,
and a mechanical stirrer is flushed with nitrogen and charged
with 200 g. (2.18 moles) of freshly distilled norbornadiene
(Note 2), 400 ml. of dry toluene, and 7.8 g. of cobalt bromide-
triphenylphosphine catalyst (Note 3). While stirring at room
temperature, 2.1 ml. of boron trifluoride etherate cocatalyst
(Note 4) is added dropwise. The mixture is then heated slowly
to 105° when the heating mantle is lowered. The ensuing
exothermic reaction maintains the temperature at 105–110° for
15 minutes. When the temperature begins to fall, the mantle
is raised, the mixture is brought to the reflux temperature,
and stirring and refluxing are continued for 12 hours. The cooled

mixture is diluted with 650 ml. of dichloromethane, transferred
to a separatory funnel, and washed with three 650-ml. portions
of water. The organic phase is dried over anhydrous magnesium
sulfate, and the solvents are evaporated at reduced pressure.
The residual crude material, 185–203 g., is distilled at 106–107°
(1.5 mm.) to give 165–170 g. (82–85%) of Binor-S which, on
cooling, solidifies to a white solid, m.p. 59–63°.

B. *Tetrahydro-Binor-S*. Binor-S (135.0 g., 0.73 mole) is
dissolved in 670 ml. of glacial acetic acid containing 5.7 ml. of
concentrated hydrochloric acid. To this solution is added 1.0 g.
of platinum oxide catalyst. The reaction mixture is hydrogenated at 200 p.s.i. hydrogen pressure and 70° for 3 hours using
a 1200-ml. glass-lined autoclave (Note 5). After cooling to room
temperature, the catalyst is removed by suction filtration
and water is added to the filtrate until two layers form. About
1.5 l. of water is required. The bottom layer, containing only
tetrahydro-Binor-S, is removed, and the top layer, consisting of
a mixture of acetic acid and water, is extracted with 400 ml.
and then two 100-ml. portions of dichloromethane. The combined dichloromethane-tetrahydro-Binor-S layers are washed
twice with 100 ml. of water, dried over anhydrous magnesium
sulfate and then evaporated under reduced pressure. The
residual tetrahydro-Binor-S is purified by distillation under
reduced pressure, b.p. 105–110° (1.5 mm.), to give 125–130 g.
(90–94%) of colorless liquid (Note 6).

C. *Diamantane*. . A 500-ml. three-necked flask, equipped
with a reflux condenser, a drying tube, a magnetic stirring bar,
and a dropping funnel is charged with 28 g. (0.1 mole) of fresh
aluminum bromide and 100 ml. of cyclohexane (Note 7). The
apparatus is flushed with hydrogen bromide gas (Note 8). When
the aluminum bromide has dissolved, 100 g. (0.53 mole) of
hydrogenated Binor-S is added dropwise to the rapidly stirred
solution, and the reaction mixture refluxes for a short time
without external heat. The course of the reaction is monitored
by g.l.c. until no more starting material remains (Note 9).
Occasionally, an additional 5 g. portion of aluminum bromide,
and application of external heat are needed to complete the
reaction. The total reaction time is about 2–3 hours (Note 10).

The hot cyclohexane layer is carefully decanted, and the aluminum bromide layer is extracted with five 200-ml. portions of hot cyclohexane. Ether (400 ml.) is added to the cooled cyclohexane extracts (Note 11), and the combined solvent fractions are washed with two 100-ml. portions of water and dried over anhydrous magnesium sulfate. Evaporation of the solvent leaves a semi-solid residue that is partially dissolved in about 100 ml. of pentane. The undissolved white solid, diamantane, is collected by suction filtration. Additional diamantane is obtained by concentrating the pentane solution to a small volume and collecting the solid that precipitates. The total amount of diamantane obtained, after drying, is 60–62 g. (60–62%), m.p. 240–241° (closed tube) (Note 12). This product is sufficiently pure for most purposes, but it may be purified further by recrystallization from pentane to give white crystals, m.p. 244.0–245.4°.

2. Notes

1. Teflon sleeves are useful to keep the joints from freezing.

2. Once distilled, the norbornadiene may be stored below 0°. Samples as old as 2 weeks were used successfully.

3. This catalyst[3] is prepared in quantitative yield by refluxing 200 ml. of benzene solution containing 10 g. (0.046 mole) of anhydrous cobalt dibromide and 24.4 g. (0.092 mole) of triphenyl phosphine. A color change is observed, and the blue-green solid that precipitates on cooling to room temperature is filtered and dried. This catalyst, stored in a dry atmosphere, appears to be active indefinitely.

4. Boron trifluoride etherate may be used without prior distillation only if fresh material is available. Care must be taken with this reagent because of fuming. The dimerization does not proceed without this cocatalyst.

5. At 70° Binor-S remains in solution, and the uptake of hydrogen is rapid. The checkers have observed that occasionally hydrogen uptake is incomplete, and an additional 1 g. of catalyst must be added to complete the absorption of hydrogen. The submitters carried out the hydrogenation in a large-scale

Parr apparatus under three atmospheres of pressure with similar results.

6. After a small solvent-containing forefraction, which is discarded, essentially all of the material should distill in the indicated range, but occasionally material boiling as high as 130° (1.5 mm.) is obtained. This is included in the product.

7. Either carbon disulfide or cyclohexane may be used with comparable yields. The advantage of carbon disulfide is the greater solubility of diamantane. When extracting with cyclohexane, a boiling solution must be used to increase solubility. However, cyclohexane is less poisonous, does not have a foul odor, and gives a whiter product. Therefore, the use of carbon disulfide was not examined by the checkers.

8. Flushing the apparatus with hydrogen bromide may not be necessary, especially for small-scale runs.

9. A Carbowax 20M or 1500 gas chromatography column at a temperature of 180° may be used. Diamantane has a shorter retention time than tetrahydro-Binor-S. The checkers used a 10% OV101 GCQ column, 100/120, at 200°; retention times are Binor-S, 11.5 minutes; tetrahydro-Binor-S, 7.6 minutes; diamantane, 6.2 minutes.

10. About 30 minutes after the addition of tetrahydro-Binor-S is complete, the reaction mixture begins to cool and external heat must be supplied to continue the refluxing and complete the reaction. If g.l.c. monitoring reveals that the rearrangement is proceeding slowly, an additional 0.5 g. of aluminum bromide is added and refluxing is continued until all the starting material is converted to product.

11. Addition of ether prevents crystallization of the diamantane from cooled cyclohexane.

12. The material obtained is pure by g.l.c. and n.m.r. The n.m.r. spectrum of diamantane shows only a singlet at $\delta 1.68$ ($CDCl_3$).[4] The pentane mother liquors contain a by-product. A comparable yield can be obtained using aluminum chloride in boiling dichloromethane, and the crude mixture at the end of the reaction need not, in some instances, even be worked up before subsequent functionalization reactions are carried out.[2,5]

3. Discussion

Like adamantane,[6] diamantane (also known as congressane[4]), the second member of the diamond family, may also be prepared by aluminum halide-catalyzed isomerization. A variety of starting materials have been shown to give diamantane,[2,4,7,8] but the very best results are obtained by the present procedure[2,8] starting with Binor-S.[2,3,9] Diamantane may be converted to a variety of functionalized derivatives.[2,5,10,11] 4-Methyldiamantane may also be prepared by rearrangement.[10]

1. Department of Chemistry, Princeton University, Princeton, New Jersey 08540. This work was supported by Grant GM-19134, National Institutes of Health, and by Hoffmann-La Roche Inc., Nutley, New Jersey 07110.
2. T. Courtney, D. E. Johnston, M. A. McKervey, and J. J. Rooney, *J. Chem. Soc., Perkins Trans. I*, 2691 (1972).
3. G. N. Schrauzer, R. K. Y. Ho, and G. Schlesinger, *Tetrahedron Lett.*, 543 (1970).
4. C. Cupas, P. v. R. Schleyer, and D. J. Trecker, *J. Amer. Chem. Soc.*, **87**, 917 (1965).
5. D. Faulkner, R. A. Glendinning, D. E. Johnston, and M. A. McKervey, *Tetrahedron Lett.*, 1671 (1971).
6. P. v. R. Schleyer, M. M. Donaldson, R. D. Nicholas, and C. Cupas, *Org. Syn.*, **42**, 8 (1962).
7. V. Z. Williams, Jr., P. v. R. Schleyer, G. J. Gleicher, and L. B. Rodewald, *J. Amer. Chem. Soc.*, **88**, 3862 (1966).
8. T. M. Gund, V. Z. Williams, Jr., E. Osawa, and P. v. R. Schleyer, *Tetrahedron Lett.*, 3877 (1970).
9. G. N. Schrauzer, B. N. Bastian, and G. A. Fosselius, *J. Amer. Chem. Soc.*, **88**, 4890 (1966).
10. T. M. Gund, M. Nomura, V. Z. Williams, Jr., and P. v. R. Schleyer, *Tetrahedron Lett.*, 4875 (1970); R. Hamilton, D. E. Johnston, M. A. McKervey, and J. J. Rooney, *Chem. Commun.*, 1209 (1972).
11. T. M. Gund and P. v. R. Schleyer, *Tetrahedron Lett.*, 1583 (1971).

DIAZOACETOPHENONE

(Acetophenone, 2-diazo)

$$C_6H_5COCl + CH_2N_2 + (C_2H_5)_3N \longrightarrow$$
$$C_6H_5COCHN_2 + (C_2H_5)_3\overset{+}{N}H\overset{-}{Cl}$$

Submitted by JOHN N. BRIDSON and JOHN HOOZ[1]
Checked by DENNIS R. MURAYAMA and RONALD BRESLOW

1. Procedure

Caution! All operations should be conducted in an efficient fume hood. Diazomethane is hazardous; directions for safe handling are given in earlier volumes of Organic Syntheses.[2,3] Diazoacetophenone is a skin irritant and direct contact should be avoided.

A solution of 0.375 mole of diazomethane in 1 l. of ether (Note 1) is placed in a 2-l. flask that is fitted with a large magnetic stirring bar and a two-necked adapter equipped with a drying tube (containing KOH pellets) and a pressure-equalizing dropping funnel. Triethylamine (37.9 g., 52.1 ml., 0.375 mole) (Note 2) is added, and the flask contents are cooled to *ca.* −10° to −5°. To the stirred mixture is added a solution of 52.75 g. (43.6 ml., 0.375 mole) of benzoyl chloride (Note 3) in 300 ml. of dry ether over a period of 0.5 hour (Note 4). An additional 50 ml. of ether is rinsed through the dropping funnel. Stirring is continued for one hour at approximately 0°, then overnight at room temperature.

The resulting precipitate of triethylamine hydrochloride (41.4 g., 81%) is filtered and washed with 100 ml. of dry ether. The solvent is removed from the combined filtrate by rotary evaporation under reduced pressure. The semi-solid residue crystallizes to an orange-red solid after refrigeration for several hours at *ca.* 5°. Crystallization from a mixture of 150 ml. of pentane and 120 ml. of dry ether affords 38.8 g. of diazoacetophenone as yellow square plates, m.p. 44–48°. Concentration of the mother liquor and extraction of the residue with boiling pentane yields an additional 7.8 g. of pale yellow rods, m.p. 47.5–48.5°, bringing the total yield to 46.6 g. (85%) (Notes 5 and 6).

2. Notes

1. Diazomethane was prepared by the method of Moore and Reed,[3] using 10% extra 2-(2-ethoxyethoxy)ethanol and an extra 100 ml. of water over that recommended to prevent stirring difficulties in the later stages of the distillation. The ethereal diazomethane solution was dried at 0° over KOH pellets, and the concentration was determined by reaction of an aliquot with benzoic acid and determining the resulting methyl benzoate by vapor phase chromatography.

2. Triethylamine, purchased from J. T. Baker Chemical Company, was refluxed over calcium hydride, then fractionally distilled through a 40-cm. Vigreux column, b.p. 81–82° (700 mm.); b.p. 89.5–90° (760 mm.).

3. The benzoyl chloride, obtained from British Drug House (Canada) Ltd., was purified[4] by washing a benzene solution with 5% aqueous sodium bicarbonate solution, drying over calcium chloride, and fractional distillation through a 40-cm. Vigreux column, b.p. 69–71° (12 mm.). The checkers used a fresh bottle, from Matheson Coleman and Bell, without purification.

4. In the later stages of the addition, a cake of crystals forms preventing adequate stirring. This difficulty is overcome by temporarily interrupting the addition and swirling the flask manually—stirring then continues normally.

5. The submitters obtained a similar yield on twice the scale reported here.

6. Although crystallization from pentane gives better crystals with an improved melting point range, recrystallization of the whole batch would require approximately 3 l. of solvent. Samples obtained from both ether-pentane and pentane evolve the theoretical amount of nitrogen on titration with $3N$ hydrochloric acid.

3. Discussion

Apart from the reaction of diazomethane with benzoyl chloride,[5,6] diazoacetophenone has been prepared by the reaction of 2-aminoacetophenone hydrochloride with sodium nitrite,[7] from the mixed anhydride of benzoic acid and ethyl carbonate with diazomethane,[8] from benzoyl chloride and potassium methyldiazotate,[9] by treating the enamine formed from 2-formylacetophenone and N-methylaniline with p-toluenesulfonyl azide,[10] and from the reaction of the sodium enolate of 2-formylacetophenone with p-toluenesulfonyl azide.[11]

The reaction of an acid chloride with diazomethane illustrates a general method of preparing diazoketones. The acid chloride is slowly added to at least two equivalents of diazomethane; the hydrogen chloride liberated (eq. 1) is then consumed according to eq. 2. When the order of addition is reversed (*e.g.*, acid chloride is in excess) and only 1 mole of diazomethane is employed, the diazoketone reacts with hydrogen chloride (eq. 3) to form the α-chloroketone.

$$RCOCl + CH_2N_2 \longrightarrow RCOCHN_2 + HCl \qquad (1)$$

$$CH_2N_2 + HCl \longrightarrow CH_3Cl + N_2 \qquad (2)$$

$$RCOCHN_2 + HCl \longrightarrow RCOCH_2Cl + N_2 \qquad (3)$$

The method described here, discovered by Newman and Beal,[6] employs triethylamine (1 equivalent) to react with the hydrogen chloride; thus only one equivalent of diazomethane is necessary. This modification is restricted to the preparation of aromatic diazoketones–aliphatic acid chlorides give a mixture of products.[6]

1. Department of Chemistry, University of Alberta, Edmonton 7, Alberta, Canada.
2. Th. J. deBoer and H. J. Backer, *Org. Syn.*, Coll. Vol. **4**, 250 (1963).
3. J. A. Moore and D. E. Reed, *Org. Syn.*, **41**, 16 (1961).
4. T. S. Oakwood and C. A. Weisgerber, *Org. Syn.*, Coll. Vol. **3**, 112 (1955).
5. F. Arndt and J. Amende, *Ber.*, **61**, 1122 (1928).
6. M. S. Newman and P. Beal III, *J. Amer. Chem. Soc.*, **71**, 1506 (1949).
7. H. E. Baumgarten and C. H. Anderson, *J. Amer. Chem. Soc.*, **83**, 399 (1961).
8. D. S. Tarbell and J. A. Price, *J. Org. Chem.*, **22**, 245 (1957).

9. E. Mueller, W. Hoppe, H. Hagenmaier, H. Haiss, R. Huber, W. Rundel, and H. Suhr, *Chem. Ber.*, **96**, 1712 (1963).
10. R. Fusco, G. Bianchetti, D. Pocar, and R. Ugo, *Chem. Ber.*, **96**, 802 (1963).
11. M. Regitz, F. Menz, and J. Rueter, *Tetrahedron Lett.*, 739 (1967).

DIDEUTERIODIAZOMETHANE

(Methane-d_2, diazo)

$$CH_2N_2 \xrightarrow[\substack{\text{tetrahydrofuran,} \\ \text{ether, O}^\circ}]{\text{NaOD, D}_2\text{O}} CD_2N_2$$

Submitted by P. G. GASSMAN[1] and W. J. GREENLEE
Checked by DAVID G. MELILLO and HERBERT O. HOUSE

1. Procedure

Caution! Diazomethane is toxic and explosive. The operations described in this procedure must be carried out in a good hood with an adequate shield (Note 1).

A distilled ethereal solution (300 ml.) containing approximately 0.96 mole of diazomethane (Note 1) is prepared from 22.5 g. of a 70% dispersion (15.8 g., 0.063 mole) of bis-(N-methyl-N-nitroso)terephthalamide (Note 2), 75 ml. of aqueous 30% sodium hydroxide, 55 ml. of diethylene glycol monoethyl ether, and 375 ml. of ether by the procedure described in an earlier volume of this series.[5] The receiving flask containing the ethereal diazomethane is capped with a rubber stopper fitted with a drying tube containing potassium hydroxide pellets to protect the solution from atmospheric moisture. The concentration of diazomethane in this ether solution may be determined either by titration with ethereal benzoic acid (Note 3) or spectrophotometrically (Note 4).

In a dry 250-ml. Erlenmeyer flask equipped with a Teflon-coated magnetic stirring bar is placed 11 ml. of a solution (Note 5) containing 0.01 mole of sodium deuteroxide in 10 ml. of deuterium oxide and 1 ml. of anhydrous tetrahydrofuran.

After the solution has been cooled in an ice bath, 120 ml. of the ethereal solution containing 0.039 mole of diazomethane is added, the flask is stoppered loosely with a cork, and the reaction mixture is stirred vigorously at 0° for one hour. The lower deuterium oxide layer is removed with a pipette and a fresh 11-ml. portion of the sodium deuteroxide solution is added. This mixture is then stirred for one hour at 0°, and the process is repeated until a total of four exchanges have been performed. The ethereal diazomethane solution is then decanted into a clean, dry 250-ml. Erlenmeyer flask and dried over 10 g. of anhydrous sodium carbonate. The resulting solution (approximately 110 ml.) contains (spectrophotometric analysis, Note 4, or titration with benzoic acid, Note 3) 0.020–0.022 mole (51–56%) of dideuteriodiazomethane which is 98–99% deuterated (Note 6).

2. Notes

1. Diazomethane is not only toxic but also potentially explosive. Hence one should wear heavy gloves and goggles and work behind a safety screen or a hood door with safety glass as is recommended in the preparation of diazomethane described by deBoer and Backer.[2] As is also recommended there, ground joints and sharp surfaces should be avoided. Thus all glass tubes should be carefully fire-polished, connections should be made with rubber stoppers, and separatory funnels should be avoided, as should etched or scratched flasks. Explosion of diazomethane has been observed at the moment crystals (sharp edges!) suddenly separated from a supersaturated solution. Stirring by means of a Teflon-coated magnetic stirrer is greatly to be preferred to swirling the reaction mixture by hand (there has been at least one case of a chemist whose hand was injured by an explosion during the preparation of diazomethane in a hand-swirled reaction vessel). It is imperative that diazomethane solutions not be exposed to direct sunlight or placed near a strong artificial light because light is thought to have been responsible for some of the explosions encountered with diazomethane. Particular caution

should be exercised when an organic solvent boiling higher than ether is used. Because such a solvent has a vapor pressure lower than ether, the concentration of diazomethane in the vapor above the reaction mixture is greater and an explosion is more apt to occur. Since most diazomethane explosions occur during distillation, procedures that avoid distillation offer certain advantages. An ether solution of diazomethane satisfactory for many uses can be prepared as described by Arndt,[3] where nitrosomethylurea is added to a mixture of ether and 50% aqueous potassium hydroxide and the ether solution of diazomethane is subsequently decanted from the aqueous layer and dried over potassium hydroxide pellets (not sharp-edged sticks!). However, the reported potent carcinogenicity[4] of nitrosomethylurea mitigates other advantages of this procedure. Two procedures involving distillation of diazomethane, those of deBoer and Backer[2] and Moore and Reed,[5] can be recommended. In neither case is there much diazomethane present in the distilling flask. The hazards associated with diazomethane are discussed by Gutsche.[6]

2. The submitters used an undistilled ethereal solution of diazomethane, prepared from nitrosomethylurea (Note 1).[3] For use in the hydrogen-deuterium exchange reaction described, ethereal diazomethane solutions prepared by any standard preparative procedures (Note 1)[2,3,5] appear to be equally satisfactory.

3. The concentration of diazomethane may be determined by reaction of an aliquot of the ethereal solution with a weighed excess of benzoic acid in cold (0°) ether solution as described in an earlier volume of this series.[3] The unchanged benzoic acid is then determined by titration with standard aqueous $0.1M$ potassium hydroxide.

4. *Caution! The following spectrophotometric analysis should be performed in a hood.* To determine the concentration of diazomethane in the ether solution obtained in this preparation, a 5-ml. aliquot of the distilled solution is diluted to 25 ml. with ether, and a portion of this solution is placed in a cylindrical Pyrex cell with an internal diameter of 1.0 cm. The optical density of the solution is determined at 410 mμ with

a suitable colorimeter such as a Bausch and Lomb Spectronic 20. From the molecular extinction coefficient, ε 7.2, at 410 mμ for diazomethane in ether solution, the concentration of diazomethane can be calculated. In a typical preparation the optical density of the diluted solution at 410 mμ was 0.46 corresponding to a diazomethane concentration of 0.064M; thus the concentration of the undiluted solution was 0.32M, corresponding to a 77% yield of diazomethane.

5. It is convenient to prepare 110 ml. of this solution at a time. Since hydrogen is evolved, the solution should be prepared in a hood. A dry 250-ml. three-necked flask is fitted with a magnetic stirrer, a rubber septum, a glass stopper, and a 125-ml. Erlenmeyer flask attached to the third neck of the reaction flask with a 10-cm. length of Gooch rubber tubing or nylon tubing. The apparatus is flushed with nitrogen from a hypodermic needle inserted through the rubber septum. Small, freshly cut chips of metallic sodium (2.3 g., 0.10 g.-atom) are placed in the Erlenmeyer flask and 100 ml. of deuterium oxide (99.7% pure grade obtained from Columbia Organic Chemicals Company, Inc.,) is placed in the reaction flask. With a hypodermic needle inserted through the rubber septum to permit the escape of hydrogen, the sodium chips are added, slowly and with stirring, to the reaction vessel. When reaction with the sodium is complete, the solution is diluted with 10 ml. of anhydrous tetrahydrofuran and stored under a nitrogen atmosphere.

6. The deuterium content can be determined by reaction of the deuterated diazomethane with benzoic acid-O-d in anhydrous ether followed by analysis of methyl benzoate for deuterium content either by n.m.r. spectrometry or by mass spectrometry. The benzoic acid-O-d is prepared by heating a mixture of 48.6 g. (0.216 mole) of benzoic anhydride (obtained from Aldrich Chemical Company, Inc.), 0.10 g. (0.0009 mole) of anhydrous sodium carbonate, and 7.0 g. (0.35 mole) of deuterium oxide to 90° for 2 hours. The resulting mixture is distilled at atmospheric pressure in a short-path still fitted with a receiver protected from atmospheric moisture by a drying tube. After removal of a forerun, b.p. 100–101°, the benzoic

acid-O-d is collected at 245–247°. During the distillation it is necessary to warm the distillation apparatus with a heat gun or an infrared lamp to prevent solidification of the benzoic acid-O-d before it reaches the receiver.

Caution! *The following reaction should be performed in a good hood* (*Note 1*). A cold (0°) solution of 1.43 g. (0.0116 mole) of benzoic acid-O-d in 10 ml. of anhydrous ether is placed in a dry 100-ml. round-bottomed flask fitted with a rubber stopper and a Teflon-coated magnetic stirring bar. The flask is cooled in an ice bath, and a sufficient amount of the ethereal dideuteriodiazomethane solution is added from a pipette to provide excess dideuteriodiazomethane in the reaction mixture. The reaction flask is stoppered loosely, and the resulting yellow solution is stirred at 0° for 10 minutes and then concentrated by first warming the solution on a steam bath in the hood and then removing the last traces of solvent under reduced pressure. The residual liquid methyl benzoate (1.4–1.5 g., 90–95% yield) is then analyzed for deuterium content. For an n.m.r. analysis, the spectrum of the pure liquid is taken and the extent of deuteration is determined by integration of the areas under the multiplet in the region 7.1–8.3 p.p.m. (aromatic CH) and the peak at 3.82 p.p.m. (OCH_3). For mass spectrometric analysis, the mass spectra of the deuterated sample and a sample of undeuterated methyl benzoate each are measured at an ionizing potential sufficiently low (approximately 12 eV.) to minimize the formation of an M-1 fragment at m/e 135 in the spectrum of the undeuterated sample. The relative abundances of the m/e 136 and 137 peaks in the spectrum of the undeuterated sample are then used to correct the peaks at m/e 137, 138, and 139 in the deuterated sample for contributions from the ^{13}C isotope. From the relative abundances of the m/e 136 peak and the corrected m/e 137, 138, and 139 peaks in the spectrum of the deuterated sample the relative proportions of d_0, d_1, d_2, and d_3, species in the deuterated methyl benzoate can be calculated. Both n.m.r. and mass spectral analysis indicated the methyl benzoate to be 98% deuterated (6–7% d_2 species and 93–94% d_3 species). When the dideuteriodiazomethane solution was allowed to react with undeuterated benzoic acid, hydrogen-deuterium exchange occurred more rapidly than esterification. The methyl benzoate produced was 70% deuterated (n.m.r. analysis) and contained 4% d_0, 37% d_1, 30% d_2, and 29% d_3 species (mass spectral analysis).

3. Discussion

The exchange procedure described was developed in the submitter's laboratories for the preparation of dideuteriodiazomethane for labeling studies. It is basically a modification of a procedure that has been used extensively in the literature.[7] However, the literature procedures give relatively little detail. This modified procedure permits the synthesis of fairly large amounts of high-purity dideuteriodiazomethane. Dideuteriodiazomethane has also been prepared from N-nitrosomethyl-d_3-urea and related trideuterated diazomethane precursors.[8] Deuterated chloroform and hydrazine hydrate have also been used to prepare dideuteriodiazomethane.[9]

The procedure described provides a general method for the hydrogen-deuterium exchange of simple diazoalkanes.

1. Department of Chemistry, The Ohio State University, 140 West 18th Avenue, Columbus, Ohio 43210.
2. Th. J. deBoer, and H. J. Backer, *Org. Syn.*, Coll. Vol. **4**, 250 (1963).
3. F. Arndt, *Org. Syn.*, Coll. Vol. **2**, 165 (1943).
4. A. Graffi and F. Hoffman, *Acta Biol. Med. Ger.*, **16**, K-1 (1966).
5. J. A. Moore and D. E. Reed, *Org. Syn.*, **41**, 16 (1961).
6. C. D. Gutsche, *Org. React.*, **8**, 391–394 (1954).
7. L. C. Leitch, P. E. Gagnon, and A. Cambron, *Can. J. Res.*, **28B**, 256 (1950); G. W. Robinson and M. McCarty, Jr., *J. Amer. Chem. Soc.*, **82**, 1859 (1960); T. D. Goldfarb and G. C. Pimentel, *J. Amer. Chem. Soc.*, **82**, 1865 (1960); H. Dahn, A. Donzel, A. Merbach, and H. Gold, *Helv. Chim. Acta.* **46**, 994 (1963); S. M. Hecht and J. W. Kozarich, *Tetrahedron Lett.*, 1501 (1972).
8. L. C. Leitch, P. E. Gagnon, and A. Cambron, *Can. J. Res.*, **28B**, 256 (1950); B. L. Crawford, Jr., W. H. Fletcher, and D. A. Ramsay, *J. Chem. Phys.*, **19**, 406 (1951); D. E. Milligan, and M. E. Jacox, *J. Chem. Phys.*, **36**, 2911 (1962); C. B. Moore and G. C. Pimentel, *J. Chem. Phys.*, **40**, 329 (1964); A. J. Merer, *Can. J. Phys.*, **42**, 1242 (1964).
9. S. P. McManus, J. T. Carroll, and C. L. Dodson, *J. Org. Chem.*, **33**, 4272 (1968).

DIETHYL 2-(CYCLOHEXYLAMINO)VINYLPHOSPHONATE

(Phosphonic acid, [2-(cyclohexylamino)vinyl]-diethyl ester)

A. $BrCH_2CH(OC_2H_5)_2 + (C_2H_5O)_3P \xrightarrow{\Delta}$

$$(C_2H_5O)_2\overset{\overset{O}{\uparrow}}{P}CH_2CH(OC_2H_5)_2 + C_2H_5Br$$

B. $(C_2H_5O)_2\overset{\overset{O}{\uparrow}}{P}CH_2CH(OC_2H_5)_2 \xrightarrow{H_3O^+}$

$$(C_2H_5O)_2\overset{\overset{O}{\uparrow}}{P}CH_2CHO + 2C_2H_5OH$$

C. $(C_2H_5O)_2\overset{\overset{O}{\uparrow}}{P}CH_2CHO + $ cyclohexyl$-NH_2 \longrightarrow$

$$(C_2H_5O)_2\overset{\overset{O}{\uparrow}}{P}CH=CH-\overset{H}{N}-\text{cyclohexyl} + H_2O$$

Submitted by WATARU NAGATA,[1] TOSHIO WAKABAYASHI,[1] and YOSHIO HAYASE[2]
Checked by H. A. DAVIS, K. ABE, and S. MASAMUNE

1. Procedure

A. *Diethyl 2,2-Diethoxyethylphosphonate* (Note 1).   Into a 2-l., three-necked round-bottomed flask fitted with a magnetic stirrer, dropping funnel, and nitrogen inlet is placed 410 g. (2.08 moles) of bromoacetaldehyde diethyl acetal (Note 2), and a gentle stream of nitrogen is then passed continuously through the system. To the stirred solution is added dropwise 316 g. (1.90 moles) of triethyl phosphite (Note 3) over a period of 30 minutes, at 110–120°. The mixture is then stirred for 3 hours at 160°. The ethyl bromide evolved is trapped by a condenser and a receiver cooled in an ice bath. The low-boiling

material (below 100°) is distilled under reduced pressure
(22 mm.). The residual oil is fractionated under reduced pressure,
and the fraction boiling at 101–103° (0.8 mm.) is collected;
yield 338 g. (70%) (Note 4).

B. *Diethyl Formylmethylphosphonate* (Note 5). A mixture
of 192 g. (0.755 mole) of diethyl 2,2-diethoxyethylphosphonate
and 670 ml. of 2% aqueous hydrochloric acid is refluxed for
10 minutes under a nitrogen atmosphere.

To the cooled (20–30°) mixture is added 240 g. of sodium
chloride (Note 6). The resulting mixture is extracted with three
500-ml. portions of methylene chloride. The combined organic
extracts are first washed with 40 ml. of 5% aqueous sodium
bicarbonate solution (Note 7), and then with 300 ml. of saturated aqueous salt solution, dried over anhydrous sodium sulfate,
and the solvent is distilled under reduced pressure (35 mm.)
on a water bath (60–70°). The resulting residue weighing
approximately 125 g. is fractionated under reduced pressure,
and the fraction boiling at 100–103° (0.8 mm.) is collected;
yield 104 g. (76%) (Notes 8 and 9).

C. *Diethyl 2-(Cyclohexylamino)vinylphosphonate.* Into a 1-l.
two-necked round-bottomed flask fitted with a magnetic
stirrer, dropping funnel, and nitrogen inlet are placed 90.0 g.
(0.50 mole) of diethyl formylmethylphosphonate and 400 ml.
of dry methanol. Under a nitrogen atmosphere, 49.6 g. (0.50
mole) of cyclohexylamine (Note 10) is added in portions to the
stirred solution, over a period of 5 minutes. During the addition
the temperature is maintained at 0–5° by cooling with an ice
bath. The mixture is stirred for an additional 10 minutes at
room temperature, and then the methanol is distilled from the
mixture under reduced pressure (10–35 mm.) on a water bath
(25–30°). The residue is dissolved in 300 ml. of dry ether
(Note 11), and dried over anhydrous potassium carbonate
(70 g.) (Note 12). After the solution is evaporated to dryness,
the residual oil is fractionated under reduced pressure in the
presence of 300 mg. of anhydrous potassium carbonate (Note
13), and the fraction boiling at 126–141° (0.08 mm.) is collected;
yield 89 g. (68%) (Note 14). The fraction above is crystallized
from dry pentane (Note 15) to yield 79 g. (60%) of diethyl
2-(cyclohexylamino)vinylphosphonate, m.p. 58–61° (Note 16).

2. Notes

1. This procedure is essentially the same as that described by Dawson and Burger.[3]

2. Bromoacetaldehyde diethyl acetal is distilled; b.p. 74–76° (25 mm.). Its preparation is described in *Org. Syn.* [Coll. Vol. **3**, 123 (1955)].

3. Triethyl phosphite is distilled; b.p. 43–44° (10 mm.). Its preparation is described in *Org. Syn.* [Coll. Vol. **4**, 955 (1963)].

4. The reported[3] boiling point is 128–130° (2 mm.). The checkers found the product to be pure by gas chromatographic analysis (UCW-98 at 150°). The n.m.r. spectrum had the following characteristic peaks (deuteriochloroform solution, internal tetramethylsilane reference): δ 1.22, 1.34(2t, 12, $J = 7$ Hz., $4CH_3$), 2.17(dd, 2, $J = 19$ and 6 Hz., CH_2CH), 3.6[2q, 4, $J = 7$Hz., $CH(OCH_2CH_3)_2$], 4.1[2q, 4, $J = 7$ Hz., $(CH_3CH_2O)_2PO$], 4.90 [dt, 1, $J = 6$ and ≈ 6 Hz., $CH_2CH(OC_2H_5)_2$].

5. This procedure is essentially the same as that described by Dawson and Burger.[3]

6. It is necessary to saturate the aqueous layer with sodium chloride in order to extract the product effectively.

7. The checkers found that it was necessary to neutralize any excess acid to prevent polymerization of the product during the distillation.

8. There is no report of the boiling point of the product in the literature.[3] The product has an infrared absorption (CCl_4 solution) at 1732 cm.$^{-1}$ (aldehyde $C{=}O$).

9. The checkers found that the product contained a 5–6% contamination of starting material [gas chromatography (UCW-98, 120°) and n.m.r.]. The n.m.r. spectrum (deuteriochloroform solution, tetramethylsilane reference): δ 1.35(t, 6, $J = 7$ Hz., 2 CH_3), 3.11[dd, 2, $J = 22$ and 3.5 Hz., $P(O)CH_2CH$], 4.2[2q, 4, $J = 7$ Hz., $(CH_3CH_2O)_2PO$], 9.70 (dt, 1, $J = 3$ and 1 Hz., CHO).

10. Reagent grade cyclohexylamine was redistilled before use.

11. Dry ether was used to minimize the water content of the solution.

12. To avoid acid-induced dimerization[4] of diethyl 2-

(cyclohexylamino)vinylphosphonate, the solution was dried thoroughly, usually overnight, with anhydrous potassium carbonate.

13. The distillation was carried out in the presence of powdered anhydrous potassium carbonate to prevent dimerization[4] of diethyl 2-(cyclohexylamino)vinylphosphonate.

14. The yields were 65–75% in several runs. The distilled oil was sufficiently pure for further use.

15. Reagent grade pentane passed through Merck anhydrous neutral alumina was used. Crystallization was carried out at 0° using 50 ml. of dry pentane. Filtration of the crystals was carried out under dry nitrogen. The submitters succeeded in crystallizing diethyl 2-(cyclohexylamino)vinylphosphonate only after the publication of its preparation.[4] The crystalline product is stable for several months, if stored at 0° under anhydrous conditions.

16. The checkers were not able to obtain the product in crystalline form. The n.m.r. spectrum (deuteriochloroform solution, tetramethylsilane reference): δ 1.3 (t, $J = 7$ Hz., CH_3CH_2O), 1.0–2.1 (m, cyclohexyl), 4.0 (quintuplet, $J = 7$ Hz., 2 CH_3CH_2O), 4.3–5.0 (broad, NH), 5.9–7.7 (m, —CH=CH—). The number of olefinic protons was estimated to be 1.7–1.8 by comparison of the area in the region δ 5.9–7.7 with the total area in the spectrum. Although this value is slightly low, it was found that the sample was sufficiently pure to carry out further transformations. The infrared spectrum has the following absorptions (neat): 3300 (N—H str.), 1620, 1210, 1058, 1035, and 955 cm.$^{-1}$.

3.   Discussion

Diethyl 2-(cyclohexylamino)vinylphosphonate has proved to be an excellent reagent for conversion of aldehydes and ketones into the corresponding α,β-unsaturated aldehydes.[4,5] Preparation of formylmethylenetriphenylphosphorane[6] and diethyl 2,2-(diethoxy)ethylphosphonate[7] have been reported and used for the conversion of aldehydes, but not ketones and hindered aldehydes, into α,β-unsaturated aldehydes. When ethyl diethyl-phosphonoacetate[8] or diethyl cyanomethylphosphonate[9] is used to obtain an α,β-unsaturated aldehyde, the conversion requires several stages.

1. Shionogi Research Laboratory, Shionogi & Co., Ltd., Fukushima-ku, Osaka, 553 Japan.
2. Faculty of Science, Osaka City University, Sumiyoshi-ku, Osaka, 558 Japan.
3. N. D. Dawson and A. Burger, *J. Amer. Chem. Soc.*, **74**, 5312 (1952).
4. W. Nagata and Y. Hayase, *Tetrahedron Lett.*, 4359 (1968); *J. Chem. Soc.*, *C*, 460 (1969).
5. W. Nagata, T. Wakabayashi, and Y. Hayase, *Org. Syn.*, **53**, 44 (1973).
6. S. Trippett and D. M. Walker, *J. Chem. Soc.*, 1266 (1961).
7. H. Takahashi, K. Fujiwara, and M. Ohta, *Bull. Chem. Soc. Jap.*, **35**, 1498 (1962).
8. W. S. Wadsworth and W. D. Emmons, *J. Amer. Chem. Soc.*, **83**, 1733 (1961).
9. A. K. Bose and R. T. Dahill, Jr., *J. Org. Chem.*, **30**, 505 (1965).

4,4-DIMETHYL-2-CYCLOHEXEN-1-ONE

$$(CH_3)_2CHCHO + \underset{H}{N}\text{-pyrrolidine} \xrightarrow[-H_2O]{\text{reflux}} (CH_3)_2C{=}CH{-}N$$

$$\xrightarrow[25°]{CH_2{=}CHCOCH_3} \quad \text{(dihydropyran intermediate)} \quad \xrightarrow[25°]{HCl, H_2O} (CH_3)_2\text{-cyclohexenone}{=}O$$

Submitted by Yihlin Chan and William W. Epstein[1]
Checked by Michael J. Umen and Herbert O. House

1. Procedure

A. *1-(2-Methylpropenyl)pyrrolidine.* A 200-ml. three-necked flask is equipped with a magnetic stirring bar, a heating mantle, a pressure-equalizing dropping funnel, a glass stopper, and a continuous water separator (a Dean-Stark trap, Note 1) fitted with a condenser and a nitrogen inlet tube. The reaction vessel is flushed with nitrogen and 61.5 g. (0.853 mole) of isobutyraldehyde (Note 2) is added to the reaction flask. An additional amount of isobutyraldehyde (Note 2) is added to the continuous water separator to fill the water-collecting trap. A static nitrogen atmosphere is maintained in the reaction vessel throughout the subsequent reaction and distillation. To the reaction flask is added, dropwise and with stirring during 5 minutes, 60.6 g. (0.852 mole) of pyrrolidine (Note 3). After this addition is complete, the dropping funnel is replaced with a glass stopper and the reaction mixture is refluxed with stirring for 3.5 hours during which time about 15 ml. (0.83 mole) of water collects in the water separator (Note 4). The water separator and condenser are replaced with a distillation head, and the reaction mixture is distilled under reduced pressure. The enamine product is collected as 99.1–100.7 g. (94–95%) of colorless liquid, b.p. 92–106° (115–118 mm.), n^{25}D 1.4708–1.4738 (Note 5).

B. *4,4-Dimethyl-2-cyclohexen-1-one.* A dry 1-l. three-necked flask is equipped with a mechanical stirrer, a pressure-equalizing dropping funnel, a nitrogen inlet tube, and an ice-water cooling bath. After the apparatus has been flushed with nitrogen, a static nitrogen atmosphere is maintained in the reaction vessel throughout the reaction. The 1-(2-methyl-propenyl)pyrrolidine (62.6 g. or 0.501 mole) is added to the reaction flask, and then 42.1 g. (0.601 mole) of methyl vinyl ketone (Note 6) is added, dropwise with stirring and cooling, during 5 minutes. After the resulting mixture has been stirred with cooling for 10 minutes, the ice bath is removed and stirring at room temperature is continued for 4 hours (Note 7). The reaction mixture is again cooled with an ice-water bath, and then 250 ml. of aqueous $8M$ hydrochloric acid is added, dropwise and with stirring (Note 8). After this addition is complete, the mixture is stirred with cooling for 10 minutes and then stirred at room temperature for 14 hours (Note 9). The resulting brown reaction mixture is extracted with two 300-ml. portions of ether, and the residual aqueous phase is neutralized by the cautious addition of 150–155 g. of solid sodium bicarbonate. This neutral aqueous phase is extracted with two 400-ml. portions of ether, and the combined ethereal extracts (Note 10) are dried over anhydrous sodium sulfate and concentrated with a rotary evaporator. The residual liquid is distilled under reduced pressure to separate 44.2–53.0 g. (71–85%) of 4,4-dimethyl-2-cyclohexen-1-one as a colorless liquid, b.p. 73–74° (14 mm.), n^{25}D 1.4699–1.4726 (Note 11).

2.   Notes

1. An illustration of a continuous water separator is provided in an earlier volume of *Organic Syntheses*.[2]

2. The checkers employed isobutyraldehyde from Eastman Organic Chemicals. The aldehyde, b.p. 62–63°, was freshly distilled from a few milligrams of *p*-toluenesulfonic acid.

3. The pyrrolidine, obtained from Aldrich Chemical Company, Inc., was redistilled before use; b.p. 88–89°.

4. The water should not be drained from the water separator during the course of the reaction.

5. The enamine has infrared absorption (pure liquid) at 1676 cm.$^{-1}$ (enamine C=C).

6. Methyl vinyl ketone (b.p. 35–36° at 140 mm.), obtained from Aldrich Chemical Company, Inc., was distilled immediately before use.

7. The use of solvents such as anhydrous ether or benzene is not only unnecessary but also undesirable, since yields are decreased by their presence. For best results the Diels-Alder adduct, which has been characterized by Opitz and Holtmann,[3] should not be isolated for the subsequent hydrolysis and cyclization.

8. The hydrolysis product, 2,2-dimethyl-5-oxo-hexanal, can be isolated if desired by stirring the Diels-Alder adduct with either aqueous 50% acetic acid or aqueous 2M hydrochloric acid followed by extraction with ether and distillation;[4] b.p. 92–94° (20 mm.).

9. Cyclization can also be accomplished by the use of an ion exchange resin.[5] On a 0.1-mole scale, 110 ml. of aqueous 1M hydrochloric acid and 70 ml. of wet Amberlite 1R-120 resin (acidified with hydrochloric acid) are added to the Diels-Alder adduct, and the mixture is refluxed for 24 hours. The

mixture is cooled and washed with four portions of ether. The ether extract is dried over anhydrous magnesium sulfate and distilled to separate 9.7–10.8 g. (78–87%) of 4,4-dimethyl-2-cyclohexen-1-one; b.p. 73-74° (14 mm.).

10. Washing the ethereal extract with either dilute acid, aqueous sodium bicarbonate, or saturated brine only decreased the yield of product and therefore is omitted.

11. The product exhibits a single peak (retention time 5.7 minutes) on a 4-m. gas chromatography column packed with silicone fluid QF_1 on Chromosorb P and heated to 191°. This material has the following spectral characteristics: i.r. (CCl_4) 1675 (conjugated C=O) and 1623 cm.$^{-1}$ (conjugated C=C); u.v. (95% C_2H_5OH) max 224 (10,600) and 321 mμ (34); n.m.r. (CCl_4) δ 1.17 (s, 6, $2CH_3$), 1.7–2.6 (m, 4, $2CH_2$), 5.71 (d, 1, $J = 10$ Hz., vinyl CH), and 6.65 (d of t, 1, $J = 10$ and 1.5 Hz., vinyl CH); m/e (rel. int.), 124(M^+, 49), 96(83), 82(100), 81(56), 68(25), 67(42), 53(22), 43(21), 41(25), and 39(25).

3. Discussion

The procedure described is essentially that of Opitz and Holtmann.[3] The yield has been increased from 27% to 85% by making changes as indicated in Notes 7 and 10. 4, 4-Dimethyl-2-cyclohexen-1-one has also been prepared by Michael addition of methyl vinyl ketone to isobutyraldehyde followed by ring formation in basic media with yields of 25%,[6] 35%,[7] and 43%.[8] This procedure has general utility in preparing 4-substituted or 4,4-disubstituted cyclohexen-2-ones and as such constitutes a useful substitute for the Robinson annelation reactions.[9,10] Unlike the latter, the alkylation step in this procedure does not require strongly basic conditions. Consequently, side reactions such as aldol condensation of the carbonyl compounds and polymerization of methyl vinyl ketone are avoided. Moreover, since the location of the double bond in the enamines is controlled by the amines used,[11] it may be possible to direct the alkylation to either side of a given carbonyl group making this procedure potentially very versatile. Similar reactions using acrolein and enamines of cyclic ketones have been utilized for the synthesis of bicyclo[n.3.1] systems.[9,12]

1. Department of Chemistry, University of Utah, Salt Lake City, Utah, 84112.
2. M. Renoll and M. S. Newman, *Org. Syn.*, Coll. Vol. **3**, 502 (1955).
3. G. Opitz and H. Holtmann, *Justus Liebigs Ann. Chem.*, **684**, 79 (1965).
4. See also I. Flemming and M. H. Karger, *J. Chem. Soc. C*, 226 (1967).
5. R. D. Allan, B. G. Cordiner, and R. J. Wells, *Tetrahedron Lett.*, 6055 (1968).
6. E. L. Eliel and C. A. Lukach, *J. Amer. Chem. Soc.*, **79**, 5986 (1957).
7. J. M. Conia and A. Le Craz, *Bull. Soc. Chim. Fr.*, 1934 (1960).
8. E. D. Bergmann and R. Corett, *J. Org. Chem.*, **23**, 1507 (1958).
9. G. Stork, A. Brizzolara, H. Landesman, J. Szmuszkovicz, and R. Terrell, *J. Amer. Chem. Soc.*, **85**, 207 (1963).
10. E. D. Bergmann, D. Ginsburg, and R. Pappo, *Org. React.*, **10**, 179 (1959).
11. W. D. Gurowitz and M. A. Joseph, *J. Org. Chem.*, **32**, 3289 (1967).
12. Y. Chan, Ph.D. Dissertation, University of Utah, 1971; A. C. Cope, D. L. Nealy, P. Scheiner, and G. Wood, *J. Amer. Chem. Soc.*, **87**, 3130 (1965).

3,5-DINITROBENZALDEHYDE

(Benzaldehyde, 3,5-dinitro-)

$$\text{(3,5-(O}_2\text{N)}_2\text{C}_6\text{H}_3\text{COCl)} + \text{LiAlH[OC(CH}_3)_3]_3 \xrightarrow[-78°]{\text{diglyme}} \text{(3,5-(O}_2\text{N)}_2\text{C}_6\text{H}_3\text{CHO)} + \text{LiCl} + \text{Al[OC(CH}_3)_3]_3$$

Submitted by J. E. Siggins,[1] A. A. Larsen,[2]
J. H. Ackerman,[1] and C. D. Carabateas[1]
Checked by D. J. Bichan and Peter Yates

1. Procedure

A 3-l. three-necked round-bottomed flask is equipped with an efficient stirrer, a pressure-equalizing dropping funnel with a nitrogen inlet, and a Y-tube fitted with a low temperature thermometer and a nitrogen outlet. The outlet is vented through a bubbler tube to maintain a slight positive pressure. The flask and dropping funnel are flamed in a stream of dry nitrogen (Note 1). To the flask is added 115.0 g. (0.50 mole) of 3,5-dinitrobenzoyl chloride (Note 2) followed by 500 ml. of dry diglyme (Note 3). The solution is stirred vigorously, and the flask is immersed in a cooling bath at $-78°$ (Note 4). A diglyme solution of lithium aluminum tri-*tert*-butoxyhydride (Note 5) is prepared in the following manner. Dry diglyme (450 ml.) is added with vigorous stirring to an Erlenmeyer flask containing 140.0 g. (0.55 mole) of lithium aluminum tri-*tert*-butoxyhydride. After standing overnight, the resulting suspension is filtered under a blanket of dry nitrogen through a thick layer of Celite packed tightly on a Buchner funnel (Note 6). The flask containing the filtrate is kept stoppered until the reducing agent is transferred to the dropping funnel. Dropwise addition of this solution is commenced when the contents of the reaction flask reach $-72°$. There is a color change and a temperature rise of a few degrees. The rate of addition is adjusted to maintain the temperature of the mixture between $-78°$ and $-68°$ (Note 7). After addition is complete the mixture is stirred at $-78°$ for 30 minutes longer.

The cold reaction mixture is poured slowly with stirring into a 3-l. beaker containing 150 ml. of concentrated hydrochloric acid, 300 ml. of saturated aqueous sodium chloride, and 150 g. of ice. A white precipitate starts to separate (Note 8). An additional 150 ml. of saturated aqueous sodium chloride is added to the beaker and, after a minute, an upper layer begins to appear. The contents are transferred to a 2-l. separatory funnel and allowed to stand for 15 to 30 minutes while an upper brown layer separates. The upper layer is reserved while the lower layer is extracted with several portions of benzene totalling 900 ml. The upper layer and the benzene extracts are combined and washed seven times with 1-l. portions of water containing 10 ml. of concentrated hydrochloric acid. The benzene layer is then washed successively with 100-ml. portions of aqueous 2% sodium bicarbonate until the washings are basic. It is dried over 100 g. of anhydrous sodium sulfate, treated with 1 g. of charcoal, and filtered. The filtrate is concentrated at reduced pressure to give 59–62 g. (60–63%) of crude 3,5-dinitrobenzaldehyde, as a tan solid, m.p. 76–80°. Trituration in an ice bath with cold dry ether (*ca.* 0.3 ml./g.) gives a spongy solid, m.p. 85–87° (lit. 85°),[3] with losses of 5–10%. This is sufficiently pure for most uses. Further purification may be effected by recrystallization from toluene-hexane.

2. Notes

1. These operations are best done the day before the experiment is performed.

2. Since commercial 3,5-dinitrobenzoyl chloride is of low purity because of contamination with 3,5-dinitrobenzoic acid, it was treated with thionyl chloride in boiling benzene under dry nitrogen. The product obtained after evaporation under vacuum melted at 68–70° and was stored over phosphorus pentoxide and potassium hydroxide in a vacuum desiccator.

3. "Diglyme" is the dimethyl ether of diethylene glycol. Commercial diglyme (Ansul Ether 141, Ansul Chemical Company, Marinette, Wisconsin) was used after drying over lithium aluminum hydride followed by distillation under reduced pressure; b.p. 59–61° (15 mm.).

4. An insulated bucket such as the "Nicer" available from B.F. Goodrich Company contains the mixture of dry ice and isopropyl alcohol. Acetone foams excessively and has a high vapor pressure.

5. This hydride is obtained from Ventron, Inc., Beverly, Massachusetts.

6. Suspended particles will plug the dropping funnel in the subsequent operation. Two funnels may be used if the filtration is too slow.

7. At elevated temperatures overreduction to the alcohol takes place. The addition time varies from 75 to 100 minutes.

8. The supernatant liquid is a brilliant yellow. A troublesome blue color may appear in an occasional run.

3. Discussion

3,5-Dinitrobenzaldehyde has been made previously by reducing 4-bromo-3,5-dinitrobenzaldehyde with cuprous hydride.[3]

The method herein described is that of H. C. Brown.[4] Lithium aluminum tri-*tert*-butoxyhydride reduction of acid chlorides to aldehydes is a synthetic method of wide utility and the yields of 20 aromatic and 10 aliphatic aldehydes so prepared have been tabulated.[5] This reagent is also used widely to reduce steroid aldehydes and ketones to alcohols, frequently at 0°.[6] Having only one active hydrogen, it has been used for the partial reduction of diketones to hydroxy ketones.[7] With proper temperature control, it does not affect oxido,[8] ester,[6,9] acetal,[10] nitrile, or nitro groups, or lactone rings,[8] all of which react with lithium aluminum hydride. In contrast to some other complex hydrides, this reagent may reduce a ketone stereoselectively to give a high relative yield of the more stable epimeric alcohol[11] and an improved absolute yield.[7,12]

1. Sterling-Winthrop Research Institute, Rensselaer, New York.
2. Mead Johnson Research Center, Evansville, Indiana.
3. H. H. Hodgson and E. W. Smith, *J. Chem. Soc. London*, 315 (1933).
4. H. C. Brown and R. F. McFarlin, *J. Amer. Chem. Soc.*, **80**, 5372 (1958).
5. H. C. Brown and B. C. Subba Rao, *J. Amer. Chem. Soc.*, **80**, 5377 (1958).
6. K. Heusler, P. Wieland, and C. Meystre, *Org. Syn.*, **45**, 57 (1965).
7. R. E. Ireland and J. A. Marshall, *J. Org. Chem.*, **27**, 1620 (1962).
8. C. Tamm, *Helv. Chim. Acta*, **43**, 338 (1960).
9. A. Bowers, E. Denot, L. C. Ibáñez, M. E. Cabezas, and H. J. Ringold, *J. Org. Chem.*, **27**, 1862 (1962).
10. J. A. Zderic and J. Iriarte, *J. Org. Chem.*, **27**, 1756 (1962).
11. J. Fajkoš, *Collect. Czech. Chem. Commun.*, **24**, 2284 (1959).
12. O. H. Wheeler and J. L. Mateos, *Chem. Ind.* (*London*), 395 (1957).

DIRECTED LITHIATION OF AROMATIC COMPOUNDS: (2-DIMETHYLAMINO-5-METHYLPHENYL)DIPHENYLCARBINOL

[Methanol, (2-dimethylamino-5-methylphenyl)diphenyl]

Submitted by J. V. Hay and T. M. Harris[1]
Checked by Robert A. Auerbach and Herbert O. House

1. Procedure

A dry 500-ml. two-necked flask containing 6.75 g. (0.05 mole) of N,N-dimethyl-p-toluidine (Note 1) and 175 ml. of anhydrous hexane (Note 2) is fitted with a Teflon-coated magnetic stirring bar, a pressure-equalizing dropping funnel capped with a rubber septum, and a nitrogen inlet tube. The reaction vessel is flushed with nitrogen and a static nitrogen atmosphere is maintained within the apparatus for the remainder of the reaction sequence. A solution of 8.8 g. (0.076 mole of N,N,N′,N′-tetramethylethylenediamine (Note 3) in 40 ml. of anhydrous hexane is added to the dropping funnel, and then a hexane solution containing 0.076 mole of n-butyllithium (Note 4) is added to the funnel. The resulting solution, which becomes warm as the organolithium-diamine complex forms, is allowed to stand for 15 minutes, and then added to the reaction mixture, dropwise and with stirring during 15–20 minutes.

The resulting bright yellow, turbid reaction mixture is stirred at room temperature for 4 hours longer, and then a solution of 13.8 g. (0.076 mole) of benzophenone (Note 5) in 40 ml. of anhydrous ether is added to the reaction mixture, dropwise and with stirring during 20 minutes. The resulting deep-green solution is stirred for an additional 20 minutes, and then poured into a vigorously stirred solution of 12 g. (0.2 mole) of acetic acid in 30 ml. of ether (Note 6). After the reaction solution has been successively extracted with 50 ml. of water and with four 50-ml. portions of aqueous 5% hydrochloric acid, the aqueous extracts are combined (Note 7) and made basic with aqueous 10% sodium hydroxide. The alkaline aqueous mixture is heated to boiling and maintained at this temperature until the escaping vapor is no longer basic to moistened pHydrion paper (Note 8). The mixture is then cooled and the white solid product which separates is collected on a Buchner funnel and washed with three 20-ml. portions of water. The crude product (m.p. 142–168°) is recrystallized from 250 ml. of a hexane-ethyl acetate mixture (3:1, v/v) to give 6.6–8.2 g. of the amino alcohol product as colorless prisms, m.p. 168–171°. Concentration of the mother liquors gives an additional 0.8–1.2 g. of product, m.p. 167–169°. The total yield is 7.8–9.0 g. (49–57%). Although the product is sufficiently pure for most purposes, a second recrystallization from a hexane-ethyl acetate mixture raises the melting point to 169.5–172° (Note 9).

2. Notes

1. Commercial N,N-dimethyl-*p*-toluidine, obtained from Aldrich Chemical Company, Inc., was used without purification.

2. An A.C.S. grade of hexane, obtained from Fisher Scientific Company, was used without further purification.

3. N,N,N′,N′,-Tetramethylethylenediamine, purchased from Aldrich Chemical Company, Inc., was distilled from calcium hydride immediately before use; b.p. 120–122°.

4. Hexane solutions of *n*-butyllithium, purchased from either Alfa Inorganics, Inc. or Foote Mineral Company, were

standardized by the titration procedure of Watson and East-ham.[2] A detailed procedure for this titration is provided in an earlier volume of *Organic Syntheses*.[3]

5. Benzophenone, purchased from Fisher Scientific Company, was used without purification.

6. Reversal of this hydrolysis procedure, the addition of acetic acid to the reaction mixture, had an adverse effect on the yield of product.

7. Some of the aqueous extracts may contain small amounts of suspended particulate matter.

8. This simple steam distillation removes the unchanged N,N-dimethyl-*p*-toluidine present in the crude product.

9. The purified product has the following spectral properties: i.r. (CCl$_4$) 3050 cm.$^{-1}$ (associated OH); u.v. (95% C$_2$H$_5$OH) max 251 (905), 258 (823), 264 (720), and 275 mμ (ε 432); n.m.r. (CDCl$_3$) δ 2.19 (s, 3, C—CH_3), 2.38 [s, 6, N(CH_3)$_2$] 6.55 (s, 1, OH), and 7.00–7.50 (m, 13, aromatic CH); m/e (rel. int.), 317(M^+, 100), 240(84), 225(28), 224(91), 222(43), 150(25), 134(41), 120(32), 105(41), 91(51), and 77(42).

3. Discussion

This procedure is an adaptation of one described by Hauser and co-workers.[4] The product has also been prepared from 2-bromo-N,N-dimethyl-*p*-toluidine by halogen-metal interchange with *n*-butyllithium followed by condensation with benzo-phenone,[4] a procedure that is less convenient than that presently described.

Tertiary amines such as N,N,N′,N′,-tetramethylethylenedia-mine (TMEDA) and 1,4-diazabicyclo[2.2.2]octane strongly catalyze metallations by alkyllithium reagents. Uncatalyzed lithiation of toluene is very poor[5] whereas by contrast, a yield of 90% has been obtained when TMEDA is employed as a catalyst.[6]

It is noteworthy that metallation of N,N-dimethyl-*p*-toluidine takes place at the position *ortho* to the dimethylamino group rather than on or *ortho* to the aryl methyl group. Apparently, coordination of lithium by the aromatic amino group plays a dominant role in directing metallation even in the presence of TMEDA. Other cases have been reported in which the site of metallation is altered by addition of TMEDA.[7,8]

1. Department of Chemistry, Vanderbilt University, Nashville, Tennessee 37235.
2. S. C. Watson and J. F. Eastham, *J. Organometal. Chem.*, **9**, 165 (1967)
3. M. Gall and H. O. House, *Org. Syn.*, **52**, 39 (1972).
4. R. E. Ludt, G. P. Crowther, and C. R. Hauser, *J. Org. Chem.*, **35**, 1288 (1970).
5. H. Gilman and B. J. Gaj, *J. Org. Chem.*, **28**, 1725 (1963).
6. A. W. Langer, Jr., *Trans. N. Y. Acad. Sci.*, **27**, 741 (1965).
7. D. A. Shirley and C. F. Cheng, *J. Organometal. Chem.*, **20**, 251 (1969).
8. D. W. Slocum, G. Book, and C. A. Jennings, *Tetrahedron Lett.*, 3443 (1970).

A GENERAL SYNTHESIS OF 4-ISOXAZOLECARBOXYLIC ESTERS: ETHYL 3-ETHYL-5-METHYL-4-ISOXAZOLECARBOXYLATE

Submitted by JOHN E. McMURRY[1]
Checked by U. P. HOCHSTRASSER and G. BÜCHI

1. Procedure

Caution! The following reactions should be performed in an efficient hood to protect the experimentalist from noxious vapors (pyrrolidine, phosphorus oxychloride, and triethylamine).

A. *Ethyl β-Pyrrolidinocrotonate.* Ethyl acetoacetate (130 g., 1.0 mole) (Note 1) and pyrrolidine (71 g., 1.0 mole) are dissolved in 400 ml. of benzene and placed in a 1-l. one-necked flask fitted with a Dean-Stark water separator on top of which is a condenser fitted with a nitrogen inlet tube. The reaction mixture is placed under a nitrogen atmosphere (Note 2) and then it is brought to and maintained at a vigorous reflux for 45 minutes, at which time the theoretical amount of water (18 ml.) has been collected. The benzene is then removed with a rotary evaporator. The residual ethyl β-pyrrolidinocrotonate is highly pure and may be used without distillation (Note 3); yield 180 g. (98%).

B. *Ethyl 3-Ethyl-5-methyl-4-isoxazolecarboxylate.* Ethyl β-pyrrolidinocrotonate (183 g., 1.0 mole), nitropropane (115 g., 116 ml., 1 l. mole), and triethylamine (400 ml.) are dissolved in 1 l. of chloroform and placed in a 5-l. three-necked flask fitted with a 500 ml. pressure-equalizing dropping funnel and a gas inlet tube. The flask is then cooled in an ice bath, and its contents are placed under a nitrogen atmosphere. While the contents of the flask are stirred magnetically, a solution prepared by dissolving phosphorus oxychloride (170 g., 1.1 mole) in 200 ml. of chloroform is placed in the dropping funnel and added slowly. After 3 hours, addition is complete and the ice bath is removed. The reaction mixture is allowed to warm to room temperature and stirring is continued for 15 hours longer.

The reaction mixture is poured into a 4-l. separatory funnel and washed with 1 l. of cold water. The chloroform layer is then washed with $6N$ aqueous hydrochloric acid until the amine bases are removed and the wash remains acid (Note 4). The chloroform extracts are washed successively with 5% aqueous sodium hydroxide (Note 5) and saturated brine, dried over anhydrous magnesium sulfate, and filtered. The solvent is removed with a rotary evaporator, and the product is distilled under vacuum to yield 122–130 g. (68–71%) of ethyl 3-ethyl-5-methyl-4-isoxazolecarboxylate, b.p. 72° (0.5 mm.), n^{23}D 1.4615 (Note 6).

2. Notes

1. The following reagents were used as supplied: pyrrolidine and triethylamine, Aldrich Chemical Company, Inc.; ethyl acetoacetate, Eastman Organic Chemicals (white label); nitropane, Matheson, Coleman and Bell (practical); phosphorus oxychloride, Matheson, Coleman and Bell (reagent).

2. Ethyl β-pyrrolidinocrotonate is typical of most enamines in that it discolors rapidly when exposed to air and therefore must be handled under an inert atmosphere.

3. Distillation is unnecessary and inadvisable, since discoloration usually occurs and there are product losses.

4. If this acid wash is not done thoroughly, the triethylamine hydrochloride which remains will sublime during distillation of the product and coat the still with a fluffy white powder. This impurity can be removed from the distillate, however, by a simple water wash.

5. This alkaline wash removes traces of ethyl acetoacetate which might form by hydrolysis of unreacted starting material during the preceding acid wash.

6. The product had the following spectral data: i.r: 1725, 1605, 1300 cm.$^{-1}$; n.m.r. (CCl_4): δ 1.3 (6H, m, 2 CH_2CH_3), 2.6 (3H, s, C=CCH_3), 2.7 (2H, q, CH_2CH$_3$), 4.2 (2H, q, CH_2CH$_3$).

3. Discussion

This procedure is illustrative of a general method for preparing a wide range of pure 3,5-disubstituted-4-isoxazole-carboxylic esters and (by hydrolysis) their acids,[2] free from positional isomers. A wide range of both primary nitro compounds and of enamino esters can be used,[2,3] and the esters thus obtained may then be used as reagents in the isoxazole annelation reaction.[3,4] The only other general synthesis of these compounds involves chloromethylation and oxidation of a suitable 4-unsubstituted isoxazole.[5] This procedure suffers from two difficulties: low yields and the unavailability of starting isoxazole. Most methods of isoxazole formation yield a mixture of position isomers,[6] but the present method is quite selective. It has been shown that the reaction of a primary nitro compound with a dehydrating agent such as phosphorous oxychloride produces an intermediate nitrile oxide[7,8] which then adds in a 1,3 dipolar cycloaddition to the enamine. This addition is remarkably selective with respect to orientation and no isomer formation is detected.[9] The intermediate isoxazoline then loses pyrrolidine to give the final product.

$$CH_3CH_2CH_2NO_2 + POCl_3 \longrightarrow CH_3CH_2C{\equiv}N{\rightarrow}O$$

1. Division of Natural Science, University of California, Santa Cruz, Santa Cruz, California 95060.
2. G. Stork and J. E. McMurry, *J. Amer. Chem. Soc.*, **89**, 5461 (1967).
3. G. Stork and J. E. McMurry, *J. Amer. Chem. Soc.*, **89**, 5464 (1967).
4. G. Stork, S. Danishevsky, and M. Ohashi, *J. Amer. Chem. Soc.*, **89**, 5459 (1967).
5. N. K. Kochetkov, E. D. Khomutova, and M. V. Bazilevskii, *J. Gen. Chem. (U.S.S.R.)*, **28**, 2762 (1958).
6. For a review of isoxazole chemistry, see; N. K. Kochetkov and S. D. Sokolov, "Advances in Heterocyclic Chemistry," Vol. 2, Academic Press, New York, 1963, pp. 365–421.
7. T. Mukaiyama and T. Hoshino, *J. Amer. Chem. Soc.*, **82**, 5339 (1960).
8. G. B. Bachman and L. E. Strom, *J. Org. Chem.*, **28**, 1150 (1963).
9. G. Bianchi, and P. Grunanger, *Tetrahedron*, **21** 817 (1965).

HOMOGENEOUS CATALYTIC HYDROGENATION: DIHYDROCARVONE

(2-Cyclohexenone, 5-isopropyl-2-methyl)

Submitted by ROBERT E. IRELAND[1] and P. BEY[2]
Checked by N. HAGA and W. NAGATA

1. Procedure

A 500-ml. two-necked creased flask, containing a magnetic stirring bar and connected to an atmospheric pressure hydrogenation apparatus equipped with a graduated burette to measure the uptake of hydrogen, is charged with 0.9 g. (0.94×10^{-3} mole) of freshly prepared tris(triphenylphosphine)-rhodium chloride (Note 1) and 160 ml. of benzene (Note 2). One neck is stoppered with a serum cap, and the mixture is stirred magnetically (Note 3) until the solution is homogeneous. The system is then evacuated and filled with hydrogen. Using a syringe, 10 g. (0.066 mole) of carvone (Note 4) is introduced into the hydrogenation flask. The syringe is rinsed with two 10-ml. portions of benzene, and the stirring is resumed. Hydrogen uptake starts immediately (Note 5) and stops 3.5 hours later when the theoretical amont of hydrogen has been absorbed. The solution is filtered through a dry column (diameter = 4 cm.) of 120 g. of Florisil (60–100 mesh). The column is washed with 300 ml. of ether and the combined solvent fractions are concentrated under reduced pressure. Vacuum distillation of the yellow residue through an 11-cm. Vigreux column (Note 6) affords 9.1–9.5 g. (90–94%) of dihydrocarvone; b.p. 100–102° (14 mm.), n^{24}D 1.479 (Notes 7 and 8).

2. Notes

1. The tris(triphenylphosphine)rhodium chloride catalyst was prepared according to the procedure of G. Wilkinson and co-workers.[4]

2. The benzene was distilled from calcium hydride.

3. Efficient stirring is necessary to assure good surface contact during the hydrogenation.

4. The carvone was distilled before use; b.p. 105–106° (14 mm.). The checkers used *l*-carvone obtained from Shiono Koryo K.K. (Japan).

5. With old catalyst, very erratic results with respect to the initiation time and the rate of hydrogen uptake have been observed.

6. When the hydrogenation is carried out on a smaller scale, purification can be affected by evaporative distillation in a bulb to bulb apparatus.

7. V.p.c. analysis shows contamination by less than 3% of carvone. The checkers used a 1 m. by 4 mm. glass column packed with 5% PEG 6000 on Chromosorb W (60/80 mesh). The retention times were 3.7 minutes and 2.75 minutes for carvone and dihydrocarvone, respectively, at 100° with a nitrogen flow rate of 90 ml. per minute.

8. The product shows the following spectral properties: i.r. (neat) 1678 cm.$^{-1}$ ($>C=C-C=O$); u.v. (ethanol) λ_{max} 237 mμ ($\varepsilon = 9150$); n.m.r. (chloroform-*d*) δ 0.88 (d, 6, $J = 6$ Hz., $(CH_3)_2CH$), 1.71 (d, 3, $J = 1.5$ Hz., $CH_3-C=CH$), 6.70 (m, 1, $HC=C$).

3. Discussion[3]

This procedure represents an example of the use of a soluble transition metal complex for the catalytic transfer of hydrogen to an olefin. First developed by Wilkinson and co-workers,[4] subsequent extensive investigation in those laboratories and others[5] has shown that the hydrogenation is sensitive to steric congestion and only unhindered double bonds are reduced. As a result, the rhodium complex has been found useful for the selective saturation of unhindered double bonds in polyolefinic substances, such as carvone.[6b] Unhindered double bonds may be reduced even in the presence of functions such as keto,[6a,b] nitro,[6b,7] and sulfide[6b] groups. The mechanism[4] and stereochemistry[8] of the catalysis due to the soluble rhodium complex have been investigated and cis-addition of hydrogen is the general rule. The catalyst is effective for deuterium addition to unhindered olefins[9] without the extensive hydrogen-deuterium exchange observed with palladium and platinum heterogeneous catalysis. The rhodium complex causes the decarbonylation of aldehydes and acid halides, and the hydrogenation of such unsaturated systems is complicated by the loss of these functional groups.[4,7] Isomerization of nonreduced olefinic bonds is also an observed side reaction.[10]

1. Division of Chemistry and Chemical Engineering, California Institute of Technology, Pasadena, California 91109.
2. Department of Chemistry, Research Laboratories, Richardson-Merrell, Inc., Strasbourg, France.
3. For further discussion and bibliography, see H. O. House, "Modern Synthetic Reactions," 2nd ed., W. A. Benjamin, Inc., New York, 1972, p. 28.
4. J. A. Osborn, F. H. Jardine, J. F. Young, and G. Wilkinson, *J. Chem. Soc. (A)*, 1711 (1966).
5. For reviews see J. Tsuji, *Advan. Org. Chem.*, **6**, 109 (1969); J. A. Osborn, *Endeavour*, **26**, 144 (1967); R. Cramer, *Accounts. Chem. Res.*, **1**, 186 (1968); R. F. Heck, *Accounts Chem. Res.*, **2**, 10 (1969).
6. (a) M. Brown and L. W. Piszkiewicz, *J. Org. Chem.*, **32**, 2013 (1967), (b) A. J. Birch and K. A. M. Walker, *J. Chem. Soc. (C)*, 1894 (1966).
7. R. E. Harmon, J. L. Parsons, D. W. Cooke, S. K. Gupta, and J. Schoolenberg, *J. Org. Chem.*, **34**, 3684 (1969).
8. F. H. Jardine, J. A. Osborn, and G. Wilkinson, *J. Chem. Soc. (A)*, 1574 (1967).
9. J. R. Morandi and H. B. Jensen, *J. Org. Chem.*, **34**, 1889 (1969).
10. J. J. Sims, V. K. Howard, and L. H. Selman, *Tetrahedron Lett.*, 87 (1969).

β-HYDROXY ESTERS FROM ETHYL ACETATE AND ALDEHYDES OR KETONES: ETHYL 1-HYDROXYCYCLOHEXYLACETATE

(Cyclohexaneacetic acid, 1-hydroxy, ethyl ester)

$$HN[Si(CH_3)_3]_2 + n\text{-}C_4H_9Li \longrightarrow LiN[Si(CH_3)_3]_2 + n\text{-}C_4H_{10}$$

$$LiN[Si(CH_3)_3]_2 + CH_3CO_2C_2H_5 \longrightarrow$$
$$LiCH_2CO_2C_2H_5 + HN[Si(CH_3)_3]_2$$

Submitted by MICHAEL W. RATHKE[1]
Checked by Y. HOYANO and S. MASAMUNE

1. Procedure

Caution! The first step of the reaction should be conducted in a well-ventilated hood since butane is liberated.

A. *Lithium Bis(trimethylsilyl)amide* (Note 1). A dry 500-ml. three-necked flask, fitted with a pressure-equalizing dropping funnel in the center neck and a stopcock in each side neck, is equipped for magnetic stirring. The flask is maintained under a static nitrogen pressure by attaching a nitrogen source to one stopcock and a mercury bubbler to the other. In the flask is placed 153 ml. of a hexane solution containing 0.250 mole of *n*-butyllithium (Note 2) and stirring is started. The flask is immersed in an ice-water bath, and 42.2 g. (0.263 mole) of hexamethyldisilazane (Note 3) is added dropwise over a period of 10 minutes. The ice-bath is removed, and the solution is stirred for 15 minutes longer. The hexane is removed under reduced pressure by replacing the mercury bubbler with heavy rubber tubing connected to a dry-ice condenser and an oil pump. During this step, the flask is immersed in a water bath at 40–50°, and stirring is continued as long as possible. After complete evaporation of the hexane, white crystals of lithium bis(trimethylsilyl)amide appear. The flask is again subjected to a static pressure of nitrogen (Note 4), and 225 ml. of tetrahydrofuran (Note 5) is added to dissolve the crystals.

B. *Lithio Ethyl Acetate.* The reaction flask is immersed in a dry ice-acetone bath, and the solution is stirred for 15 minutes to achieve temperature equilibration. After this time, 22.1 g. (0.250 mole) of ethyl acetate is added dropwise over a 10-minute period. Stirring is continued for an additional 15 minutes to complete the formation of lithio ethyl acetate (Note 6).

C. *Ethyl 1-Hydroxycyclohexylacetate.* A solution of 24.6 g. (0.250 mole) of cyclohexanone (Note 7) in 25 ml. of tetrahydrofuran is added dropwise to the reaction mixture over a 10-minute period. After an additional 5 minutes, the reaction mixture is hydrolyzed by adding 75 ml. of 20% hydrochloric acid all at once (Note 8). The cooling bath is removed, and the stirred solution is allowed to reach room temperature.

The organic layer is separated, the aqueous layer is extracted with two 50-ml. portions of ether, and the combined extracts are dried over anhydrous sodium sulfate. The solvent is removed with a rotary evaporator (Note 9), and the almost-colorless residue is distilled under reduced pressure through a 10-cm. Vigreux column to give 37–42 g. (79–90%) of the β-hydroxy ester as a colorless liquid, b.p. 77–80° (1 mm.); $n^{24}\text{D}$ 1.4555–1.4557 (Note 10).

2. Notes

1. The preparation of lithium bis(trimethylsilyl)amide is adapted from an earlier procedure.[2] The original procedure specifies addition of butyllithium to hexamethyldisilazane in ether followed by a reflux period. It is generally more convenient

to add the hexamethyldisilazane to the butyllithium and satisfactory results are obtained without using ether or refluxing.

2. A $1.63M$ solution of butyllithium in hexane was purchased from Foote Mineral Company.

3. Hexamethyldisilazane was obtained from Pierce Chemical Company and used without further purification.

4. Lithium bis(trimethylsilyl)amide is hydrolyzed rapidly by moist air. It is therefore essential to break the vacuum by admitting nitrogen rather than air.

5. The submitters used reagent grade tetrahydrofuran (available from Fisher Scientific Company) from a freshly opened bottle. The checkers used tetrahydrofuran purified by distillation from lithium aluminum hydride. See *Org. Syn.*, **46**, 105 (1966), for warning regarding purification of this solvent.

6. Solutions of lithio ethyl acetate prepared by this method are stable indefinitely at $-78°$, but decompose rapidly if allowed to reach room temperature.

7. White label cyclohexanone (Eastman Organic Chemicals) was used without further purification.

8. The yield of the β-hydroxy ester is somewhat lower if the reaction mixture is allowed to reach room temperature prior to hydrolysis.

9. Continuing the evaporation process for some time after removal of the solvent is helpful in removing any residual hexamethyldisilazane (b.p. 125°) together with its hydrolysis product, hexamethyldisiloxane (b.p. 100°).

10. N.m.r. ($CDCl_3$, internal TMS reference) δ 1.26 (t, 3, $J = 7$ Hz., CH_3), 1.52 [broad s, 10, $(CH_2)_5$], 2.45 (s, 2, CH_2COO), 3.33 (s, 1, OH), 4.19 (q, 2, $J = 7$ Hz., OCH_2).

3. Discussion

Ethyl 1-hydroxycyclohexylacetate has been prepared by the Reformatsky reaction of cyclohexanone with zinc and ethyl bromoacetate (56–71%)[3] and by the condensation of ethyl acetate with cyclohexanone in liquid ammonia using two equivalents of lithium amide (69%).[4]

This preparation illustrates a general method for the preparation of β-hydroxy esters from ethyl acetate and aldehydes or ketones.[5] The procedure is simpler and less time-consuming than other methods and the yields are usually higher. In addition, the β-hydroxy esters are obtained in a high state of purity.

The procedure is especially suited to small-scale preparations (25 mmoles or less) where the necessity of evaporating hexane from the lithium bis(trimethylsilyl)amide is much less of a handicap. In such cases, it is convenient to equip the reaction vessel with a septum inlet and transfer all reagents with a syringe.

1. Department of Chemistry, Michigan State University, East Lansing, Michigan 48823.
2. E. H. Amonoo-Neizer, R. A. Shaw, D. O. Skovlin, and B. C. Smith, *J. Chem. Soc. London*, 2997 (1965).
3. R. L. Shriner, *Org. React.*, **1**, 17 (1942).
4. W. R. Dunnavant and C. R. Hauser, *J. Org. Chem.*, **25**, 503 (1960).
5. M. W. Rathke, *J. Amer. Chem. Soc.*, **92**, 3222 (1970).

ISOXAZOLE ANNELATION REACTION:
1-METHYL-4,4a,5,6,7,8-HEXAHYDRONAPHTHALEN-2(3H)-ONE

Submitted by JOHN E. McMURRY[1]
Checked by U. P. HOCHSTRASSER and G. BÜCHI

1. Procedure

A. *3-Ethyl-4-hydroxymethyl-5-methylisoxazole. Caution! Lithium aluminum hydride can react with explosive violence on contact with water or when overheated, and great care must be taken in its handling.*

A slurry of lithium aluminum hydride (21.0 g., 0.55 mole) in anhydrous ether is prepared by cautiously adding the powdered reagent (Note 1) to 2.5 l. of freshly prepared anhydrous ether in a 5-l. three-necked flask fitted with a reflux condenser, a 500-ml. pressure-equalizing addition funnel, and a strong mechanical stirrer. The contents of the flask are then placed under a nitrogen atmosphere by means of a gas inlet tube attached to the top of the condenser. Ethyl 3-ethyl-5-methyl-isoxazolecarboxylate (124 g., 0.68 mole) (Note 2), dissolved in 300 ml. of dry ether, is placed in the addition funnel and added dropwise over 4 hours to the lithium aluminum hydride slurry (Note 3). The reaction is refluxed gently for 4 hours and then placed in an ice bath. Quenching of excess reagent and hydrolysis of aluminate salts is effected by *cautious, slow* addition of 20 ml. of water, followed by 30 ml. of 10% aqueous sodium hydroxide and another 30 ml. of water (Note 4). The ether layer is filtered from granular aluminum salts, poured into a 2-l. separatory funnel, and washed with 250 ml. of saturated brine. The organic extract is dried over anhydrous magnesium sulfate and filtered, and the solvent is removed with a rotary evaporator. The residual oil is distilled to yield 76–82 g. (80–86%) of 3-ethyl-4-hydroxymethyl-5-methylisoxazole, b.p. 99-101° (0.15 mm.); n^{23}D 1.4835; i.r.: 3450, 1640 cm.$^{-1}$; n.m.r. (CCl$_4$): δ 1.2 (3H, t, CH$_2$CH_3), 2.2 (3H, s, C=CCH_3), 2.6 (2H, q, CH_2CH$_3$), 4.2 (2H, s, CH_2O).

B. *4-Chloromethyl-3-ethyl-5-methylisoxazole. Caution! The following reaction should be carried out in a fume hood to avoid thionyl chloride vapors.*

3-Ethyl-4-hydroxymethyl-5-methylisoxazole (54 g., 0.38 mole) is dissolved in 70 ml. of methylene chloride and placed in a 500-ml. one-necked flask fitted with a 100-ml. pressure-equalizing addition funnel and a magnetic stirrer. The flask is placed in an ice bath and its contents are stirred while a solution of thionyl chloride (53 g., 32 ml., 0.45 mole) in 50 ml. of methylene chloride is added dropwise. Addition is complete in 1 hour, and the reaction is then allowed to warm to room temperature and stir for an additional hour. After retrieval of the magnetic stirring bar from the flask, the solvent is removed with a rotary evaporator and the dark residual liquid is distilled to yield 47–49 g. (78–81%) of 4-chloromethyl-3-ethyl-5-methyl-isoxazole, b.p. 77–78° (1.5 mm.); n^{23}D 1.4845; i.r.: 1620, 680 cm.$^{-1}$; n.m.r. (CCl$_4$): δ 1.3 (3H, t, CH$_2$CH_3), 2.3 (3H, s, C=CCH_3), 2.6 (2H, q, CH_2CH$_3$), 4.3 (2H, s, CH_2Cl).

C. *2-(3-Ethyl-5-methyl-4-isoxazolylmethyl)cyclohexanone.* Sodium hydride (10.0 g. of a 60% slurry in mineral oil, 0.25 mole) is degreased in a flame-dried 1-l. three-necked flask fitted with a 250-ml. pressure-equalizing addition funnel and a condenser through which a stream of nitrogen is being blown. The sodium hydride is washed by adding 20 ml. of dry benzene, stirring magnetically, allowing it to settle, and drawing off the supernatant benzene wash with a syringe. The washing process is repeated four more times and then 100 ml. of dry benzene is added, followed by 100 ml. of dry dimethylformamide (Note 5). The contents of the flask are then covered with a nitrogen atmosphere and a solution of ethyl 2-cyclohexanonecarboxylate (41.0 g., 0.25 mole) (Note 6) in 100 ml. of 1:1 benzene-dimethylformamide is added slowly over 45 minutes with cooling to keep the reaction mixture near room temperature (Note 7). A solution of 4-chloromethyl-3-ethyl-5-methylisoxazole (32 g., 0.20 mole) in 100 ml. of 1:1 benzene-dimethylformamide is then added over 30 minutes and the reaction is stirred for 2 days at room temperature. The reaction is diluted with 300 ml. of ether, poured into a 1-l. separatory funnel, washed three times with 100-ml. portions of water and once with brine, dried over anhydrous magnesium sulfate and filtered, and the solvents are removed with a rotary evaporator. The residual oil is dissolved in 150 ml. of glacial acetic acid and placed in a 500-ml. one-necked flask fitted with a magnetic stirrer and a reflux condenser. Hydrochloric acid (150 ml., $6N$) is added, the mixture is refluxed for 36 hours (Note 8) and then concentrated with a rotary evaporator. The residue is taken up in 500 ml. of ether, poured into a 1-l. separatory funnel, and washed twice with 100-ml. portions of water, once with 5% aqueous sodium hydroxide, and once with brine. After drying over anhydrous magnesium sulfate and filtration, the organic extracts are concentrated with a rotary evaporator and distilled (Note 9) to yield 33–35 g. (75–80%) of 2-(3-ethyl-5-methyl-4-isoxazolylmethyl)cyclohexanone, b.p. 130° (0.001 mm.); n^{23}D 1.4970; i.r.: 1710, 1630 cm.$^{-1}$; n.m.r. (CCl$_4$): δ 1.2 (3H, t, CH$_2$CH_3), 1.5–2.2 (11H, m, CH_2, CH), 2.3 (3H, s, C=CCH_3), 2.5 (2H, q, CH_2CH$_3$).

D. *1-Methyl-4,4a,5,6,7,8-hexahydronaphthalen-2(3H)-one.*

Caution! Sodium ethoxide formation should be carried out in a hood since a large volume of hydrogen gas is evolved.

2-(3-Ethyl-5-methyl-4-isoxazolylmethyl)cyclohexanone (27.6 g., 0.125 mole) is dissolved in 250 ml. of ethanol in a Parr hydrogenation bottle, and 20 g. of freshly prepared W-4 Raney nickel catalyst (Note 10) is added. Hydrogenation is started at an initial hydrogen pressure of 25 p.s.i. Cleavage of the isoxazole ring is complete after 6 hours, after which time the reaction is stopped and the solution is filtered free of catalyst (Note 11). The catalyst is washed with ether and with absolute ethanol, and the combined organic filtrates are concentrated with a rotary evaporator (Note 12).

The viscous residual liquid is dissolved in 25 ml. of absolute ethanol and a stream of nitrogen is bubbled through the solution for 15 minutes to remove dissolved oxygen (Note 13). A solution of sodium ethoxide is then prepared by cautiously dissolving freshly cut sodium (11.5 g., 0.500 mole) in 150 ml. of absolute ethanol under a nitrogen atmosphere in a 500-ml. three-necked flask fitted with a reflux condenser topped with a gas-inlet, a magnetic stirrer, and a rubber serum cap on one of the side-arms. When ethoxide formation is complete, the deoxygenated solution of the hydrogenated isoxazole is injected into the stirred reaction mixture through the rubber serum cap with a syringe. The solution is refluxed until the u.v. spectrum of a small aliquot withdrawn with a syringe through the serum cap shows the absence of absorption at 345 nm. (Note 14). This requires about 30 hours.

A solution of 15 ml. of glacial acetic acid and 30 ml. of water is deoxygenated as described above and then slowly injected with a syringe into the reaction. Refluxing is continued for 6 hours, the flask is cooled, and its contents are poured into a 1-l. separatory funnel along with 200 ml. of water. The solution is extracted four times with 100-ml. portions of ether, and the combined ether extracts are washed successively with 100 ml. of 6N aqueous hydrochloric acid, 100 ml. of water, and 100 ml. of brine. The organic extracts are dried over anhydrous magnesium sulfate, filtered, concentrated with a rotary evaporator, and distilled to yield 13.2–13.8 g. (65–67%) of 1-methyl-4,

4a,5,6,7,8-hexahydronaphthalen-2(3H)-one, b.p. 85–90° (0.5 mm.); n^{23}D 1.5120; i.r.: 1670, 1605 cm.$^{-1}$; n.m.r. (CCl$_4$): δ 1.7 (3H, s, C=CCH_3), 1.0–2.5 (13H, m, CH_2, CH).

2. Notes

1. The reagents used in this procedure were obtained from the following sources: lithium aluminum hydride, Alfa Inorganics, Inc.; thionyl chloride, Matheson Coleman and Bell; sodium hydride, Metal Hydrides, Inc. The nitrogen was prepurified.

2. See *Org. Syn.*, **53**, 59 (1973).

3. The addition must be done cautiously and the reaction watched constantly to see that efficient stirring is maintained. When the addition is approximately half-complete, doughy lumps, which tend to form on top of the solution, impede the stirring.

4. This quenching procedure is mentioned in L. Fieser and M. Fieser, "Reagents for Organic Synthesis," Vol. I, John Wiley and Sons, Inc., New York, 1967, p. 584.

5. The dimethylformamide was dried and purified by distillation from anhydrous copper sulfate.

6. The ethyl cyclohexanonecarboxylate was purchased from Aldrich Chemical Company, Inc. and contains approximately 40% methyl ester. The amount used takes this fact into account.

7. Dimethylformamide begins to decompose if the temperature rises too much.

8. The submitter stated that 24 hours of refluxing was sufficient for complete decarboxylation; however, the checkers found that after 24 hours at reflux, approximately 30% of the ester remained. Analysis was performed by gas chromatography (6-ft. column, 10% silicon rubber, 210°).

9. The distillation is most conveniently done in a short-path distillation apparatus with a mercury diffusion pump.

10. The W-4 Raney nickel is prepared according to A. A. Pavlic and H. Adkins, *J. Amer. Chem. Soc.*, **68**, 1471 (1946).

11. *Caution! Since the catalyst is highly pyrophoric when dry, do not suck it to dryness.*

12. The hydrogenated isoxazole is quite sensitive to air and heat and should be used as soon as possible to prevent decomposition.

13. Oxygen must be rigorously avoided, particularly in smaller scale reactions, to prevent oxidation of the dihydropyridine intermediate to the corresponding pyridine.

14. The absorption maximum at 345 nm. corresponds to an acetyldihydropyridine intermediate (see Discussion) and disappears when the acetyl group is cleaved by ethoxide. Thus the reaction can be readily followed spectroscopically.

3.   Discussion

The isoxazole annelation reaction[2,3] is a general method for fusing a new cyclohexanone ring onto an existing system and is complementary to the well-known Robinson annelation.[4] It has several major advantages:

1. The isoxazole ring serves as a "masked" or protected 3-ketobutyl side chain which can be positioned *alpha* to the existing ketone at an early stage in a complex synthesis. The isoxazole ring is stable to acids, bases, and hydride reducing agents[5] but can be cleanly and selectively cleaved by hydrogenolysis. Thus at an appropriate time, the 3-ketobutyl side chain can be unmasked and annelation completed.

2. Although the present procedure attaches the isoxazole *via* alkylation of a β-keto ester, there are several different methods by which attachment could have been effected. Both alkylation of a cyclohexanone enamine[6] and direct alkylation of an enone anion followed by hydrogenation of the enone double bond have been used successfully.[2,3]

3. Since a wide range of 3-substituted-4-chloromethylisoxazoles can be easily prepared, the isoxazole annelation sequence allows one to construct a variety of substituted cyclohexenone systems.

The mechanism of the annelation sequence is of some interest and has been shown to proceed through the following path[7]:

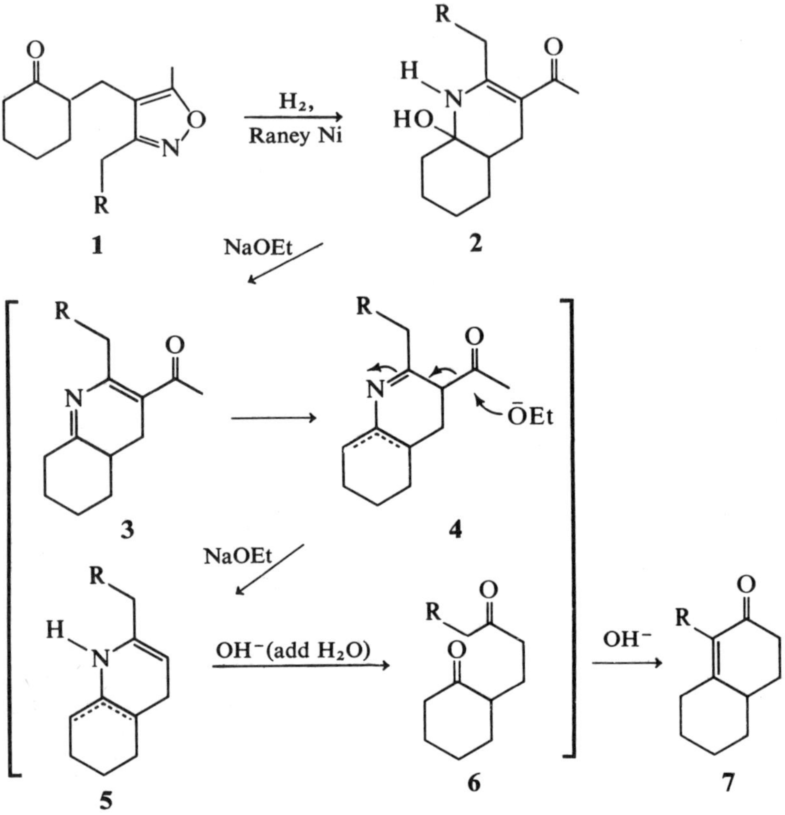

The anhydrous, deoxygenated sodium ethoxide solution readily dehydrates carbinolamide **2** to the acyldihydropyridine **3** but prevents hydrolysis or oxidation of **3**. Base-catalyzed double bond migrations can lead to **4**, the imine of a β-diketone, and the acetyl fragment can then be cleaved. Addition of water to the reaction causes hydrolysis of the cross-conjugated dienamine **5** to diketone **6** which then cyclizes.

The present procedure is illustrative of the general method which finds its utility largely in the construction of more complex polycyclic systems. The specific compound herein synthesized can be made more conveniently by standard Robinson annelation techniques.[8]

1. Division of Natural Science, University of California, Santa Cruz, Santa Cruz, California 95060.
2. G. Stork, S. Danishevsky, and M. Ohashi, *J. Amer. Chem. Soc.*, **89**, 5459 (1967).
3. G. Stork and J. E. McMurry, *J. Amer. Chem Soc.*, **89**, 5464 (1967).
4. E. C. du Feu, J. McQuillin, and R. Robinson, *J. Chem. Soc.*, 53 (1937).
5. For a review of isoxazole chemistry, see N. K. Kochetkov and D. Sokolov, "Advances in Heterocyclic Chemistry," Vol. 2, Academic Press, Inc., New York, 1963, pp. 365–421.
6. G. Stork and M. Ohashi, personal communication.
7. G. Stork and J. E. McMurry, *J. Amer. Chem. Soc.*, **89**, 5463 (1967).
8. G. Stork, A. Brizzolara, R. Landesman, J. Szmuskovicz, and R. Terrell, *J. Amer. Chem. Soc.*, **85**, 207 (1963).

KETONES AND ALCOHOLS FROM ORGANOBORANES:
1. PHENYL HEPTYL KETONE
2. 1-HEXANOL
3. 1-OCTANOL

A. $n\text{-}C_4H_9CH{=}CH_2 + BH_3 \longrightarrow (C_6H_{13})_3B$

B. $(C_6H_{13})_3B + N_2CHCOC_6H_5 \xrightarrow{H_2O} n\text{-}C_7H_{15}COC_6H_5$

C. $(C_6H_{13})_3B \xrightarrow[H_2O_2]{NaOH} n\text{-}C_6H_{13}OH + n\text{-}C_4H_9CH(OH)CH_3$

D. $(CH_3)_2C{=}CHCH_3 + BH_3 \longrightarrow [(CH_3)_2CHCH(CH_3)]_2BH$

E. $[(CH_3)_2CHCH(CH_3)]_2BH + n\text{-}C_6H_{13}CH{=}CH_2 \longrightarrow$
$$[(CH_3)_2CHCH(CH_3)]_2B{-}n\text{-}C_8H_{17}$$

F. $[(CH_3)_2CHCH(CH_3)]_2B{-}n\text{-}C_8H_{17} \xrightarrow[H_2O_2]{NaOH}$
$$(CH_3)_2CHCH(OH)CH_3 + n\text{-}C_8H_{17}OH$$

Submitted by HIROMICHI KONO and JOHN HOOZ[1]
Checked by DENNIS R. MURAYAMA, JACK EMERT, J. M. PECORARO, and RONALD BRESLOW

1. Procedure

A. *Trihexylborane.* A dry 1-l. three-necked flask is equipped with a magnetic stirring bar, a reflux condenser fitted with a drying tube, a pressure-equalizing dropping funnel to which is attached a rubber septum cap, and a three-way parallel side-arm connecting tube fitted with a thermometer and an inlet tube (containing a stopcock) to permit introduction of a dry nitrogen atmosphere. The apparatus is flushed with nitrogen and charged with 27.8 g. (0.33 mole) of 1-hexene (Note 1) followed by 150 ml. of anhydrous tetrahydrofuran (Note 2) by means of a hypodermic syringe. Then 103 ml. (0.11 mole) of a $1.07M$ solution of borane in tetrahydrofuran (Note 3) is added dropwise over a 20-minute period to the stirred solution, while the reaction temperature is maintained below approximately 20° by means of an ice bath. After the addition, the reaction mixture is stirred for an additional one hour at room temperature. The resulting solution of trihexylborane (Note 4) is ready for use in the next step.

B. *Phenyl Heptyl Ketone.* To the solution, prepared as described in Section A., is added 18 ml. (1.0 mole) of water. The nitrogen flow is ceased, and the drying tube is quickly replaced by a stopcock that is attached, by means of Tygon tubing, to a gas-measuring tube. A solution of 14.6 g. (0.10 mole) of diazoacetophenone (Note 5) in 125 ml. of tetrahydrofuran is added to the stirred solution over a period of one hour. After the addition is complete, the mixture is stirred vigorously for one hour at room temperature, then heated to reflux for one hour. The resulting mixture is cooled to approximately 25° with an ice bath (Notes 6, 7, and 8). A solution of 73 ml. (0.22 mole) of $3N$ sodium acetate solution is added, followed by the dropwise addition of 23 ml. (0.22 mole) of 30% hydrogen peroxide, while maintaining the reaction temperature below approximately 20° (ice-cooling). The cooling bath is then removed and the mixture is stirred at room temperature for one hour.

The resulting mixture is saturated with sodium chloride, and the organic phase is separated and washed with three 50-ml. portions of saturated brine solution. The organic solution is dried over sodium sulfate and concentrated on a rotary evaporator. Distillation of the residue through a 7-cm. Vigreux column separates 15.34–15.99 g. (75–80%) of phenyl heptyl ketone, b.p. 118–120° (0.60 mm.) n^{26}D 1.5034 (Note 9).

C.   *1-Hexanol.*   To a solution of trihexylborane in a 500-ml. three-necked flask [prepared from 25.3 g. (0.30 mole) of 1-hexene in 150 ml. of tetrahydrofuran and 84 ml. of a 1.20M solution of borane in tetrahydrofuran, as described in Section A] is added 34 ml. (0.10 mole) of a 3N solution of sodium hydroxide. This is followed by the dropwise addition of 36 ml. (0.35 mole) of 30% hydrogen peroxide solution at a rate such that the reaction temperature is maintained at approximately 35° (water bath). After being stirred at room temperature for one hour, the mixture is poured into 100 ml. of water. The organic phase is separated and the aqueous phase is extracted with 50 ml. of ether. The combined organic extracts are washed with three 50-ml. portions of saturated brine solution and dried over Drierite. After the bulk of the solvent is removed by distillation, the residue is fractionated with a 24-in. Teflon spinning band column (Note 10) to provide 7.7–8.2 g. (25.1–26.7%) of 1-hexanol of 95% purity, b.p. 145–153° and 10.1–15.4 g. (33.3–50.3%) of pure 1-hexanol, b.p. 153–155° (Note 11). The total yield of material with >95% purity is 58.4–77%.

D.   *Bis-(3-methyl-2-butyl)borane   ("Disiamylborane").*[5]   A dry 500-ml. three-necked flask is equipped as described in Section A. The apparatus is flushed with nitrogen, and the flask is charged with 92 ml. (0.11 mole) of a 1.2M solution of borane in tetrahydrofuran. The flask is cooled with an ice bath, and to the stirred solution is added a solution of 15.4 g. (0.22 mole) of 2-methyl-2-butene (Note 12) in 40 ml. of anhydrous tetrahydrofuran over a 30-minute period. After the addition is complete, the reaction mixture is kept below approximately 10° for 2 hours. The resulting solution is used directly in the next step.

E.   *Addition of Disiamylborane to 1-Octene.*   To the solution prepared as described in Section D is added a solution of 11.2 g. (0.10 mole) of 1-octene (Note 13) in 20 ml. of anhydrous tetrahydrofuran over a 30-minute period while the reaction temperature is maintained below approximately 20°. The ice bath is removed and stirring is continued for one hour at room temperature.

F.   *1-Octanol.*   The stirred solution prepared as described in

Section E is cooled below approximately 10° with an ice bath, and a solution of 34 ml. (0.10 mole) of $3N$ sodium hydroxide is introduced. This is followed by the dropwise addition of 36 ml. (0.3 mole) of 30% hydrogen peroxide, added at a rate such that the reaction temperature is maintained between 30–35°. After the addition is complete, the mixture is stirred at room temperature for 1.5 hours. The reaction mixture is then extracted with 100 ml. of ether. The ether extract is washed with four 100-ml. portions of water and dried over Drierite. After removal of solvent on a rotary evaporator, the residue is distilled through a 3-cm. Vigreux column to give, as a forerun, 9.6–10.5 g. of 3-methyl-2-butanol, b.p. 110–115°, followed by 8.5–9.1 g. (65–70%) of 1-octanol, b.p. 182–186° (Note 14).

2. Notes

1. The 1-hexene (99%), purchased from Aldrich Chemical Company, Inc., was stored over molecular sieves and distilled prior to use.

2. Commercial tetrahydrofuran, purchased from British Drug House (Canada) Ltd. or Fisher Scientific Company, was refluxed over sodium metal, distilled from sodium metal, then redistilled from lithium aluminum hydride under a nitrogen atmosphere.

3. A commercial one molar solution of borane in tetrahydrofuran, obtained from Alfa Inorganics, Inc., was standardized by measuring the amount of hydrogen evolved on titration with 40% aqueous ethylene glycol solution.

4. The trihexylborane solution contains approximately 94% primary and 6% secondary boron-bound alkyl groups.[2,3]

5. Diazoacetophenone was prepared as described in Ref. 4.

6. Approximately 95% of the theoretical amount of nitrogen is evolved.

7. Gas chromatography indicates a 92% yield of product. Using a 10 ft. by 0.25 in. column packed with 20% NPGSE (Neopentyl Glycol Sebacate Ester) suspended on Chromosorb W heated to 235° and a helium flow rate of 60 ml. per minute,

the submitters found a retention time of 21 minutes for phenyl heptyl ketone.

8. Distillation of the product from the crude reaction mixture at this stage gives somewhat lower yields. Therefore, residual organoboranes are oxidized prior to isolation of product.

9. Hexanol, 18.5–20.3 g. (78–86%), boiling at approximately 35° (0.75 mm.), is obtained as a forerun. The checkers found that two distillations were required to give a product of >95% purity.

10. A forerun of approximately 1.0 g., b.p. 135–145°, comprised largely of 2-hexanol (75%), is discarded. The submitters used a stainless steel spinning band column with equivalent results.

11. The product may be analyzed by use of a gas chromatography column packed with 20% SF-96 suspended on Chromosorb WAW, 5 ft. by 0.25 in., operated at 105°. Using a helium flow rate of 60 ml. per minute, the submitters found a retention time of 4 minutes.

12. Commercial 2-methyl-2-butene (99%), purchased from Chemical Samples Company, 4692 Kenny Road, Columbus, Ohio 43220, was used as received.

13. The 1-octene (97%), b.p. 122–123°, purchased from the Aldrich Chemical Company, Inc., was stored over molecular sieves and distilled prior to use.

14. The product may be analyzed by using a gas chromatography column packed with 20% SF-96 suspended on Chromosorb WAW, 5 ft. by 0.25 in., operated at 140°. Using a helium flow rate of 60 ml. per minute the submitters found a retention time of 3.5 minutes.

3. Discussion

Phenyl heptyl ketone has been prepared by the Friedel-Crafts acylation of benzene with octanoyl chloride.[6] It is also a product of the thermal decomposition of the mixed ferrous salts of benzoic and octanoic acids.[7]

The present preparation of phenyl heptyl ketone illustrates

the procedure for reaction of a trialkylborane with an α-diazo-ketone to form a homologated ketone. It is representative of a fairly general reaction between an organoborane and a stabilized diazo compound, as illustrated in the accompanying equation, to form the corresponding ketone,[8] diketone,[9] nitrile,[10] ester,[10] or aldehyde.[11]

$$R_3B + N_2CHA \quad \xrightarrow[H_2O]{-N_2} \quad RCH_2A$$

$$A = COCH_3,\ COC_6H_5,\ CO(CH_2)_nCOCHN_2,\ CN,\ CO_2C_2H_5,\ CHO$$

The extent of reaction is conveniently monitored by measuring the quantity of nitrogen evolved. Organoboranes derived from terminal olefins react readily ($>90\%$ gas evolution) at room temperature or below, whereas more highly hindered organoboranes react more sluggishly (*ca.* 3–6 hours of reflux) to complete the liberation of nitrogen.

Enol borinates are intermediates and are rapidly hydrolyzed to product in the presence of water.[11,12] Since neither the organoborane nor diazo compound reacts with water appreciably under the experimental conditions, hydrolysis is conveniently accomplished *in situ* by adding water to the solution of organoborane prior to the addition of diazo substrate. Although the product may be isolated from the crude mixture by extraction and distillation, an oxidation step (to convert residual organic boron-containing material to boric acid) is employed, since it gives somewhat higher isolated yields.

An adaptation of the procedure employing deuterium oxide as the hydrolytic medium permits the synthesis of α-deuterio ketones and esters in high isotopic purity. α,α-Dideuterio

$$R_3B + N_2CHA \quad \xrightarrow{D_2O} \quad R-\underset{\underset{D}{|}}{C}H-A$$

$$A = COCH_3,\ COC_6H_5,\ CO_2C_2H_5$$

ketones and esters are also produced in high purity using the appropriate α-deuteriodiazocarbonyl precursor.[13]

Another useful consequence of the facile *in situ* hydrolysis is the extension of the procedure to the alkylation of cyclic

$$(CH_2)_n \begin{array}{c} C=O \\ | \\ C=N_2 \end{array} \quad \xrightarrow[H_2O]{R_3B} \quad (CH_2)_n \begin{array}{c} C=O \\ | \\ HC-R \end{array}$$

α-diazo ketones[14] ($n = 3$, 4, 5, 6). This adaptation obviates the necessity of the several separate (yield-lowering) steps required for the removal of activating or blocking groups using other alkylation methods.[15]

The principal disadvantage of this procedure is that only one alkyl group of the trialkylborane is constructively utilized. The reaction is also sensitive to steric factors. Although yields are excellent for terminal olefins, reaction becomes more sluggish and yields of ketone decrease progressively as steric effects in the trialkylborane are increased. The method is thus of limited utility for rare olefins. However, the overall simplicity of procedure, mildness of reaction conditions, and absence of any isomeric contaminants recommend the method for reactions involving rarer diazocarbonyl substrates.

Apart from the oxidation of trihexylborane,[16] 1-hexanol has been prepared by a previous *Organic Syntheses*[17] procedure involving the reaction of ethylene oxide with n-butylmagnesium bromide and alternate methods of synthesis are reviewed therein.

The present preparation illustrates the procedure for the hydroboration[18] of a terminal olefin and the oxidation of the resultant trialkylborane.

The hydroboration of an olefin involves a *cis* addition of a boron-hydrogen bond to an alkene linkage, and for unsymmetric olefins occurs in a counter-Markownikoff fashion. 1-Alkenes and simple 1,2-disubstituted olefins undergo rapid conversion to the corresponding trialkylborane, whereas addition of diborane to tri- and tetrasubstituted olefins may be conveniently terminated at the respective di- and monoalkylborane stage. 1-Alkenes yield trialkylboranes in which there is a preponderant (approximately 94%) addition of the boron atom to the terminal carbon.[2,3]

The oxidation of a trialkylborane may be effected by perbenzoic acid or by aqueous hydrogen peroxide in the presence of alkali.[20] A detailed systematic study of the reaction parameters (oxidation temperature, base concentration, hydrogen peroxide concentration) of the latter method has led to the development[3] of a standard and common procedure for oxidizing organoboranes, and is illustrated in the present procedure.

The oxidation step occurs with retention of configuration of the carbon atom undergoing migration. The mechanism is believed to proceed as illustrated in the following equation.[2] As a result, the sequence involving the hydroboration of an

$$H_2O_2 + OH^- \;\rightleftharpoons\; HOO^- + H_2O$$

$$\text{>B-R} + \bar{O}OH \longrightarrow \left[\text{>B} \overset{R}{\underset{O-OH}{\cdots}} \right] \longrightarrow OH^- + \text{>B-OR}$$

$$\text{>B-OR} \xrightarrow{\text{hydrolysis}} \text{>BOH} + ROH$$

olefin followed by treatment with $NaOH$-H_2O_2 constitutes a useful device for effecting the overall counter-Markownikoff *cis* hydration of an olefin.

$$\text{(olefin)} \xrightarrow{BH_3} \text{(alkylborane)} \xrightarrow[H_2O_2]{NaOH} \text{(alcohol)}$$

The principal disadvantage of this procedure resides in its application to terminal olefins. Since the hydroboration step produces *ca.* 94% primary boron-bound alkyl groups, the maximum purity of primary carbinol is obviously limited to *ca.* 94%. Isolation of primary alcohol free of the contaminant secondary alcohol requires a tedious, yield-lowering fractionation procedure. This difficulty may be circumvented by employing a more selective hydroborating reagent, disiamylborane, as illustrated in the synthesis of 1-octanol.

1-Octanol has previously been prepared from ethyl caprylate by catalytic hydrogenolysis,[21] and by the Bouveault-Blanc method using sodium and alcohol in toluene.[22] Other preparative methods include the reaction between *n*-hexylmagnesium bromide and ethylene oxide,[23] and the oxidation of trioctyl-borane.[24]

The hydroboration of a trisubstituted olefin, exemplified by the reaction of 2-methyl-2-butene with diborane, is conveniently stopped at the dialkylborane stage to produce disiamylborane. As a result of its rather large steric requirements this

$$(CH_3)_2C{=}CHCH_3 + BH_3 \longrightarrow \left[(CH_3)_2CH{-}\overset{\overset{\displaystyle CH_3}{\displaystyle |}}{CH}{-}\right]_2 BH$$

reagent selectively hydroborates terminal olefins to place *ca.* 99% of the boron atom on the terminal carbon. Consequently, oxidation produces essentially homogeneous 1-alkanol. This procedure is the method of choice for converting terminal olefins to primary alcohols without the accompanying formation of isomers.

$$R_2BH + R'CH{=}CH_2 \longrightarrow R'{-}CH_2CH_2{-}BR_2$$

$$R = (CH_3)_2CH{-}\underset{|}{CH}(CH_3)$$

The advantages of a hydroboration-oxidation sequence to prepare alcohols are simplicity of procedure; relatively mild reaction conditions; high overall yields; absence of skeletal rearrangements; production of carbinol in which there is an overall *cis* addition of water to a double bond in a counter-Marknowikoff sense.

1. Department of Chemistry, University of Alberta, Edmonton 7, Alberta, Canada.
2. H. C. Brown, "Hydroboration," W. A. Benjamin, Inc., New York, 1962.
3. G Zweifel and H C. Brown, *Org. React.*, **13**, 1 (1963).
4. J. N. Bridson and J. Hooz, *Org. Syn.*, **53**, 35 (1973)
5. This trivial term, which now finds common usage, was coined as a contraction of the only *sec*-isoamyl structure possible, $(CH_3)_2CHCH(CH_3)$.[2]
6. F. L. Breusch and M. Oguzer, *Chem. Ber.*, **87**, 1225 (1954).
7. C. Granito and H. P. Schultz, *J. Org. Chem.*, **28**, 879 (1963).
8. J. Hooz and S. Linke, *J. Amer. Chem. Soc.*, **90**, 5936 (1968).
9. J. Hooz and D. M. Gunn, *J. Chem. Soc. (D)*, 139 (1969).
10. J. Hooz and S. Linke, *J. Amer. Chem. Soc.*, **90**, 6891 (1968).
11. J. Hooz and G. F. Morrison, *Can. J. Chem.*, **48**, 868 (1970).
12. D. J. Pasto and P. W. Wojtkowski, *Tetrahedron Lett.*, 215 (1970).
13. J. Hooz and D. M. Gunn, *J. Amer. Chem. Soc.*, **91**, 6195 (1969).

14. J. Hooz, D. M. Gunn and H. Kono, *Can J. Chem.*, **49**, 2371 (1971).
15. H. O. House, "Modern Synthetic Reactions." W. A. Benjamin, Inc., New York, 1965, p. 195.
16. H. C. Brown and G. Zweifel, *J. Amer. Chem. Soc.*, **82**, 4708 (1960).
17. E. E. Dreger, *Org. Syn.* Coll. Vol. **2**, 306 (1941).
18. Although the term hydroboration is most commonly employed[2,3,15] to denote the addition of a boron-hydrogen linkage to carbon–carbon multiple bonds, it has also been used "for the two-step oxidative process to distinguish it from the process of reduction involving H-B addition and protonolysis."[19]
19. L. F. Fieser and M. Fieser, "Reagents for Organic Synthesis," J. Wiley and Sons, Inc., New York, 1967, p. 203.
20. J. R. Johnson and M. G. Van Campen, Jr., *J. Amer. Chem. Soc.*, **60**, 121 (1938).
21. K. Folkers and H. Adkins, *J. Amer. Chem. Soc.*, **54**, 1145 (1932).
22. C. S. Marvel and A. L. Tanenbaum, *J. Amer. Chem. Soc.*, **44**, 2645 (1922).
23. R. C. Huston and A. H. Agett, *J. Org. Chem.*, **6**, 123 (1944).
24. H. C. Brown and B. C. Subba Rao, *J. Amer. Chem. Soc.*, **81**, 6423 (1959).

MODIFIED CLEMMENSEN REDUCTION: CHOLESTANE

Submitted by SHOSUKE YAMAMURA,[1] MASAAKI TODA,[2] and YOSHIMASA HIRATA[2]
Checked by A. LAURENZANO, L. A. DOLAN, and A. BROSSI

1. Procedure

A 500-ml. four-necked round-bottomed flask (Note 1) equipped with a sealed mechanical stirrer (Note 2), a gas inlet tube, a low-temperature thermometer, and a calcium chloride tube is charged with 250 ml. of dry ether. By means of a dry ice-acetone bath the temperature of the ether is regulated at -10 to $-15°$ and maintained within this range while a slow stream (Note 3) of hydrogen chloride is introduced with slow stirring for about 45 minutes. The gas inlet tube is then replaced with a glass stopper and 10.0 g. (0.026 mole) of cholestanone (Note 4) is added while the temperature of the stirred solution (Note 5) is kept below $-15°$. The reaction mixture is cooled to $-20°$ and 12.3 g. (0.19 g. atoms) of activated zinc (Note 6) is added over a 2–3 minute period. The temperature of the reaction mixture is allowed to rise to $-5°$ (Note 7), and it is maintained between $-4°$ and $0°$ (Note 8) for 2 hours. Stirring is not interrupted for the duration of the reaction. The mixture is finally cooled to $-15°$ and poured slowly onto about 130 g. of crushed ice. The ethereal layer is separated and the aqueous layer is extracted with 100 ml. of ether that had been used to rinse the reaction vessel. The ethereal solutions are combined, washed with saturated aqueous sodium chloride, dried over anhydrous magnesium sulfate, and filtered. The ether is distilled under reduced pressure with a 50° water bath to leave 9.3–9.5 g. of a colorless, liquid residue that solidifies on cooling. This solid is dissolved in 30–40 ml. of n-hexane (Note 9), poured onto a 3.5 cm. by 17 cm. column of silica gel (Note 10) and eluted with 80–90 ml. of n-hexane. Distillation of the solvent under reduced pressure with a 50° water bath leaves 8.0–8.2 g. (82–84%) of cholestane (Note 11) which, after recrystallization from ethanol-ether (Note 12), weighs 7.3–7.5 g. (76–77%); plates, m.p. 78–79° (lit. 80°)[3] (Note 13).

2. Notes

1. A standard three-necked flask fitted with a Y-tube may be used.

2. An efficient magnetic stirrer may be substituted.

3. Approximately one bubble per second can be spot-checked periodically by connecting the calcium chloride tube to an oil-filled bubble counter.

4. Prepared according to William F. Bruce, *Org. Syn.*,

Coll. Vol. **2**, 139 (1943); single spot on t.l.c. with the system described in Note 11.

5. The cholestanone does not dissolve completely at this low temperature but the reaction is not affected.

6. The submitters prepared activated zinc by the following procedure. Sixteen grams of commercial zinc powder, special grade, ca. 300 mesh, obtained from either Kishida Chemical Company Ltd. or Hayashi Pure Chemical Company Ltd., is added with stirring to a 300-ml. round-bottomed flask containing 100 ml. of 2% hydrochloric acid. Vigorous stirring is continued until the surface of the zinc becomes bright (ca. 4 minutes). The aqueous solution is decanted, and the zinc powder in the flask is washed by decantation with four 200-ml. portions of distilled water. The activated zinc powder is transferred to a suction filter with 200 ml. of distilled water and washed successively with 50 ml. of ethanol, 100 ml. of acetone, and 50 ml. of dry ether. Filtration and washing should be done as rapidly as possible to minimize exposure of the activated zinc to air. The zinc is finally dried at 85–90° for 10 minutes in a vacuum oven (ca. 15 mm.), cooled, and used immediately; yield, 13–14 g.

The checkers used this procedure with certified zinc powder, 325 mesh, obtained from Fisher Scientific Company.

7. This requires ca. 20 minutes.

8. The temperature is regulated by adding pieces of dry ice to the cooling bath as required. As the reduction proceeds, the solution separates into two phases.

9. The solution is decanted from any insoluble matter.

10. Silicic acid, 100 mesh (Mallinckrodt).

11. This material melts at 78–79°. On t.l.c. [silica, development with n-hexane, visualization with sulfuric acid-methanol (1:1) and heating] the product had $R_f = 0.74$. An impurity, $R_f = 0.65$, was present.

12. The cholestane is dissolved in 50 ml. of ether. Ether is distilled until the volume is 25 ml., 200 ml. of ethanol is added, and the mixture is refrigerated.

13. Recovery is 92%. Recrystallization has no effect on the quality of the product as judged by m.p. and t.l.c. (Note 11).

3. Discussion

The well-known Clemmensen reduction[4] is a general method by which aralkyl ketones are readily converted to the corresponding hydrocarbons with amalgamated zinc and hydrochloric acid. It is not particularly effective, however, with alicyclic and aliphatic ketones. The procedure herein described provides a simple method of reducing a variety of ketones to their desoxy derivatives in high yields under much milder conditions (0°, 1–2 hours) than those normally used in the Clemmensen reaction.[4] This permits the selective deoxygenation of ketones in polyfunctional molecules[5] containing groups such as cyano, amido, acetoxy, and carboalkoxy which are stable under the mild reaction conditions. For example, the following reduction[6] has been carried out successfully by the modification of our procedure in which acetic anhydride serves as the solvent.[5]

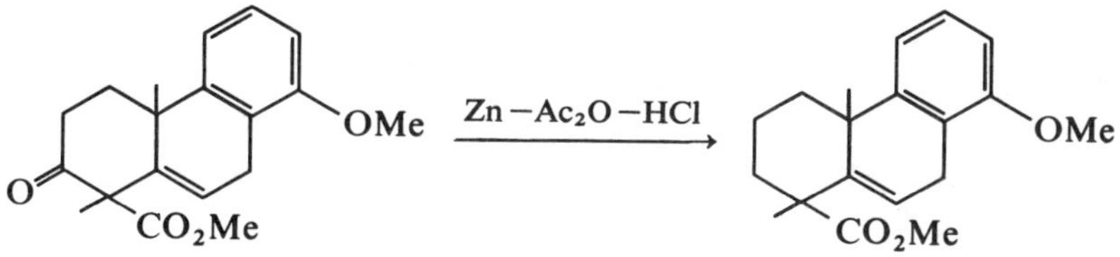

Wide latitude is permitted in choosing the solvent for the reaction. Several organic solvents (tetrahydrofuran, benzene, *n*-hexane)[7] and particularly acetic anhydride[5,8] may be used instead of dry ether. α-Halo- and α-acetoxycholestanone are converted to cholestane with Zn-HCl-Et_2O and also with Zn-HCl-Ac_2O.[7,8] These reduction systems, however, have given different results with α,β-unsaturated ketones.[9] With Zn-HCl-Et_2O, cholest-1-en-3-one gave cholestane in 88% yield while cholest-4-en-3-one gave an 88% yield of a mixture of 1.2 parts of cholestane and 1 part of coprostane. By contrast, reaction of Zn-HCl-Ac_2O with cholest-1-en-3-one afforded a mixture of three compounds: cholestane (30–32%), 3-acetoxycholest-2-ene (10–24%), and cholestanone (30–40%). Cholestanone appears to be formed from the corresponding cyclopropanol acetate[10] during the work up.

The mechanism of this reduction is probably similar to that of the Clemmensen reaction.[11]

1. Faculty of Pharmacy, Meijo University, Showa-ku, Nagoya, Japan.
2. Chemical Institute, Nagoya University, Chikusa-ku, Nagoya, Japan.
3. O. Diels, and K. Linn, *Ber.*, **41**, 548 (1908); A. Windaus, *Ber.*, **50**, 133 (1917).
4. E. L. Martin, *Org. React.*, **1**, 155 (1942).
5. S. Yamamura and Y. Hirata, *J. Chem. Soc. C*, 2887 (1968).
6. Private communication from H. Kakisawa, Tokyo Kyoiku University, Tokyo, Japan.
7. M. Toda, Y. Hirata, and S. Yamamura, *Chem. Commun.*, 919 (1969).
8. S. Yamamura, *Chem. Commun.*, 1494 (1968).
9. M. Toda, M. Hayashi, Y. Hirata, and S. Yamamura, unpublished results.
10. M. I. Elphimoff-Felkin and P. Sarda, *Tetrahedron Lett.*, 3045 (1969).
11. H. O. House, "Modern Synthetic Reactions," W. A. Benjamin, Inc., New York, 1965, p. 58; J. G. St. C. Buchanan and P. D. Woodgate, *Quart Rev. (London)*, **23**, 522 (1969) and references cited therein.

ORCINOL MONOMETHYL ETHER

(*m*-Cresol, 5-methoxy-)

Submitted by R. N. Mirrington and G. I. Feutrill[1]
Checked by H. Gurien, G. Kaplan, and A. Brossi

1. Procedure

A. *Orcinol Dimethyl Ether.* In a 1-l. three-necked flask fitted with a mechanical stirrer, a condenser, and a 100-ml. dropping funnel are placed 124 g. (0.9 mole) of anhydrous potassium carbonate, 410 ml. of acetone (Note 1), and 42.6 g. (0.3 mole) of orcinol monohydrate (Note 2). The stirrer is started and 94.5 g. (71 ml., 0.75 mole) of dimethyl sulfate is added from the dropping funnel to the pink mixture over a period of 2 minutes. The mixture warms appreciably and begins to reflux after an additional 5 minutes. When the spontaneous boiling has subsided (15–20 minutes after addition of the dimethyl sulfate), the stirred mixture is heated gently under reflux for 4 hours longer. The condenser is then arranged for distillation and 200 ml. of acetone is distilled. A 50-ml. portion of concentrated aqueous ammonia is added to the reaction mixture and stirring and heating are continued for 10 minutes. The mixture is diluted with water to a total volume of approximately 750 ml., the layers are separated, and the organic layer is combined with two 150-ml. ethereal extracts of the aqueous layer. The organic phase is washed with 50 ml. of water, twice with 50-ml. portions of $3N$ sodium hydroxide solution (Note 3), 50 ml. of saturated aqueous sodium chloride, and then dried over magnesium sulfate. After evaporation of the ether at atmospheric pressure, the residual liquid is distilled under reduced pressure to yield 42.9–43.7 g. (94–96%) of orcinol dimethyl ether, b.p. 133–135° (40 mm.) (Notes 4 and 5).

Caution! Because hydrogen is evolved and large volumes of foul-smelling ethyl methyl sulfide are liberated, this step should be conducted in a well-ventilated hood.

B.   *Orcinol Monomethyl Ether.*   In a 1-l. three-necked flask equipped with a magnetic stirrer, a condenser, a dropping funnel, and a nitrogen inlet are placed 250 ml. of dry dimethylformamide (Note 6) and 22 g. (0.55 mole) of sodium hydride (60% oil dispersion). The suspension is stirred under an atmosphere of dry nitrogen and cooled with an ice bath while a solution of 31 g. (37 ml., 0.50 mole) of ethanethiol (Note 7) in 150 ml. of dry dimethylformamide (Note 6) is added slowly from the dropping funnel over a period of 20 minutes. The ice bath is removed and stirring is continued for an additional 10 minutes. A solution of 38.0 g. (36.5 ml., 0.25 mole) of orcinol dimethyl ether in 100 ml. of dry dimethylformamide (Note 6) is added in one lot, and the mixture is refluxed under an atmosphere of dry nitrogen for 3 hours (Notes 8 and 9). The mixture is cooled, then poured into 1.8 l. of cold water, and extracted with two 250-ml. portions of petroleum ether (b.p. 50–70°) which are discarded. The aqueous layer is acidified with 330 ml. of ice-cold 4N hydrochloric acid and extracted with three 250-ml. portions of ether. The combined ethereal extracts are washed with 100 ml. of saturated aqueous sodium chloride and dried over magnesium sulfate. The ether is distilled at atmospheric pressure and the residual liquid is distilled under reduced pressure to yield 28–30.5 g. (81–88%) of orcinol monomethyl ether, b.p. 89–90° (0.2 mm.) or 156–158° (25 mm.) (Notes 10 and 11.)

2. Notes

1. Technical acetone containing about 1% water is quite satisfactory.

2. British Drug Houses Ltd. reagent grade orcinol monohydrate was used without further purification.

3. If the first washing is colorless, as is usually the case, the second washing is unnecessary. Washing with sodium hydroxide solution should be continued until the washings are colorless.

4. A similar run using 100 g. of orcinol monohydrate afforded 102 g. (95%) of orcinol dimethyl ether, b.p. 67.5–68.5° (0.2 mm.).

5. V.p.c. analysis of the product on two columns (silicone gum rubber SE-30 and OV-1) indicated the presence of traces of two other compounds with retention times longer than that of orcinol dimethyl ether. These impurities, which were most likely C-methylated materials,[2] totaled less than 0.5% of the product.

6. Dimethylformamide, b.p. 58° (25 mm.), was distilled from calcium hydride under a reduced pressure of nitrogen immediately before use.

7. British Drug Houses Ltd. reagent grade ethanethiol was distilled from calcium hydride before use (b.p. 36°).

8. The mixture may become gelatinous during this time but stirring is not necessary.

9. A polythene tube leading from the top of the condenser to the back of the hood is advisable to prevent any diffusion

of the by-product, ethyl methyl sulfide, into the laboratory. Alternatively, this by-product may be collected, if desired, by passing the vapors through a cold trap (dry ice in acetone).

10. This distillate, which is sufficiently pure for most reactions, solidifies after standing for 4–6 hours. A sample crystallizes from benzene-petroleum ether as off-white prisms, m.p. 61–62°, and is relatively free of sulfureous odor.

11. N.m.r. (CCl_4, internal TMS): δ 2.19 (s, 3, CH_3), 3.63 (s, 3, OCH_3), 6.17 (m, 3, ArH), 6.38 (broad s, 1, OH).

3. Discussion

Previous preparations of orcinol monomethyl ether have been effected by partial methylation of oricinol with methyl iodide and potassium hydroxide[3] or sodium ethoxide,[4] or with dimethyl sulfate and sodium hydroxide.[5] These procedures required tedious purification stages and the pure monomethyl ether was obtained in 37% yield at best.[5]

This procedure is characterized by the easy isolation of a high-purity product in excellent yield. The reaction illustrates a general method[6] for the conversion of aryl methyl ethers to the corresponding phenols, and has proved to be of special advantage with acid-sensitive substrates.[6,7]

A unique feature of this procedure is the selective *mono*demethylation of the dimethyl ether. The scope of this reaction is illustrated[6] in part by the preparation in high yield of *p*-methoxyphenol, guaiacol, and phloroglucinol dimethyl ether from the respective fully O-methylated compounds. An exception is pyrogallol trimethyl ether which affords pyrogallol 1-monomethyl ether in high yield.[6]

1. Department of Organic Chemistry, University of Western Australia, Nedlands, W.A., 6009.
2. D. D. Ridley, E. Ritchie, and W. C. Taylor, *Aust. J. Chem.*, **21**, 2979 (1968).
3. F. Tiemann and F. Streng, *Ber.*, **14**, 1999 (1881).
4. F. Henrich, *Monatsh. Chem.*, **22**, 232 (1901).
5. F. Henrich and G. Nachtigall, *Ber.*, **36**, 889 (1903).
6. G. I. Feutrill and R. N. Mirrington, *Tetrahedron Lett.*, 1327 (1970); *Aust. J. Chem.*, **25**, 1719 (1972).
7. G. I. Feutrill, R. N. Mirrington, and R. J. Nichols, *Aust. J. Chem.*, **26**, 345 (1973); G. I. Feutrill and R. N. Mirrington, *Aust. J. Chem.*, **26**, 357 (1973).

OXYMERCURATION-REDUCTION: ALCOHOLS FROM OLEFINS: 1-METHYLCYCLOHEXANOL

Submitted by J. M. Jerkunica and T. G. Traylor[1]
Checked by A. K. Willard and R. E. Ireland

1. Procedure

In a 3-l, three-necked flask, fitted with a thermometer and a mechanical stirrer are placed 95.7 g. (0.3 mole) of mercuric acetate (Note 1) and 300 ml. of water. After the mercuric acetate dissolves, 300 ml. of ether is added. While this suspension is stirred vigorously 28.8 g. (0.3 mole) of 1-methylcyclohexene (Notes 2 and 3) is added, and stirring is continued for 30 minutes at room temperature (Note 4). A solution of 150 ml. of $6N$ NaOH is then added followed by 300 ml. of $0.5M$ sodium borohydride solution in $3N$ NaOH. The borohydride solution is added at a rate such that the reaction mixture can be maintained at or below 25° with an ice bath.

The reaction mixture is stirred at room temperature for 2 hours after which time the mercury is found as a shiny liquid. The supernatant liquid is separated from the mercury (Note 5), the ether layer is separated and the aqueous solution is extracted with two 100-ml. portions of ether. The combined ether solutions are dried over magnesium sulfate and distilled to give 24.1–25.8 g. (70.5–75.4%) of 1-methylcyclohexanol, b.p. 154.5–156°; n^{21}D 1.4596 (Note 6).

2. Notes

1. The mercuric acetate was purchased from Mallinckrodt Chemical Works.

2. 1-Methylcyclohexene was purchased from K & K Laboratories and used without further purification.

3. Sometimes a yellow color (mercuric oxide) appears at this point and disappears as the reaction proceeds. If the yellow color does not disappear in about 10 minutes, 1.5 ml. of 70% perchloric acid per mole of mercuric acetate may be added to accelerate the reaction. Under these conditions even unreactive olefins are completely oxymercurated in about an hour.

4. The checkers found that extending the time of oxymercuration to 2 hours did not improve the yield.

5. The checkers found that the reaction mixture could be decanted from the mercury only if the mixture was allowed to stand for at least one hour after stirring was stopped. An alternate procedure which proved quite satisfactory was filtration of the entire reaction mixture through a Celite pad immediately after the stirring was stopped.

6. The distilled product slowly deposits mercury. In an effort to determine whether extent of this deposition is reduced by extending the time of reduction, the checkers found that stirring the crude alcohol with Celite for 15 hours, followed by filtration and distillation, did not diminish the amount of mercury deposited. However, in a typical run where the yield of distilled alcohol was 24.6 g. (72.0%), after standing for 24 hours, the distillate was decanted from the deposited mercury and redistilled to give 21.4 g. of 1-methylcyclohexanol which did not deposit mercury upon standing for one week at room temperature. The yield of twice-distilled alcohol was 62.6%.

3. Discussion

This method of preparing alcohols is an adaptation of an oxymercuration procedure of Sand and Genssler[2] and reduction methods of Henbest and Nicholls.[3] Other methods for preparing 1-methylcyclohexanol are oxymercuration followed by reduction in tetrahydrofuran-water;[4] reaction of cyclohexanone with methylmagnesium halides;[5] and reduction of 1-methylcyclohexene epoxide or methylenecyclohexane epoxide with lithium aluminum hydride.[6]

Although the reaction proceeds faster in tetrahydrofuran-water[4] or in acetone-water, ethyl ether was used as solvent in this reaction for convenience of product separation and purification. However, the oxymercuration is acid catalyzed,[7] and oxymercuration of unreactive olefins such as *cis*-cyclooctene can be accelerated by adding acid (Note 3). Rapid stirring also accelerates the reaction. An additional advantage of using ether is that less olefin is produced during the reduction using this solvent than when tetrahydrofuran is used. Elimination can be a serious side reaction during the reduction, amounting to 30% of total demercurated product when the oxymercurial from *cis*-cyclooctene is reduced in tetrahydrofuran-water. In ether-water, however, less than 10% olefin is produced.

As a general procedure if the olefin is impure, the oxymercuration-reduction process may include an olefin purification step. Alternatively, this process may be used to purify the olefin for other purposes.[2,4c] In such cases, acetone is substituted for ether and, after oxymercuration for the same length of time as suggested above, the solution is poured with stirring into two volumes of water containing one equivalent each of sodium bicarbonate and sodium chloride. The mercury derivative is filtered, recrystallized from ethanol-water, ether, dioxane, or ethyl acetate-heptane[8] and then either reduced as described above (in 70–80% yield) to produce pure alcohol, or deoxymercurated with cold $6N$ HCl,[2] with ethereal lithium aluminum hydride[9] (added cautiously), or high concentrations of alkali halides[4c,9,10] to produce the pure olefin.

Oxymercuration may also be used to prepare ethers, acetates, amines, or amides (Markownikoff adducts). Several excellent procedures for these syntheses have been published by H. C. Brown and co-workers.[4b]

1. Department of Chemistry, Revelle College, University of California, San Diego, LaJolla, California 92037.
2. J. Sand and O. Genssler, *Ber.*, **36**, 3705 (1903).
3. H. B. Henbest and B. Nicholls, *J. Chem. Soc. London*, 227 (1959).

4. (a) H. C. Brown and P. Geoghegan, Jr., *J. Amer. Chem. Soc.*, **89**, 1522 (1967); (b) H. C. Brown, J. H. Kawakami, and S. Misumi, *J. Org. Chem.*. **35**, 1360 (1970) (Other reduction studies are summarized here); (c) H. C. Brown and P. Geoghegan, Jr., *J. Org. Chem.*, **35**, 1844 (1970).

5. M. Barbier and M. F. Hügel, *Bull. Soc. Chim. Fr.*, 951 (1961).

6. M. Mousseron, R. Jacquier, M. Mousseron-Canet, and R. Zagdoun, *C. R. Acad. Sci., Paris*, **235**, 177 (1952).

7. W. Kitching, *Organometal. Chem. Rev.*, **3**, 61 (1968).

8. T. G. Traylor and A. W. Baker, *J. Amer. Chem. Soc.*, **85**, 2746 (1963).

9. T. G. Traylor, Thesis, University of California, Los Angeles, 1952.

10. T. G. Traylor and S. Winstein, "Abstracts of Papers," 82-0, Division of Organic Chemistry, 135th National Meeting, American Chemical Society, Boston, Mass., April 1959.

1-PHENYL-4-PHOSPHORINANONE

(4-Phosphorinanone, 1-phenyl-)

Submitted by Theodore E. Snider, Don L. Morris,
K. C. Srivastava, and K. D. Berlin[1]
Checked by John R. Berry and Richard E. Benson

1. Procedure

A. *Preparation of bis(2-Cyanoethyl)phenylphosphine.* A 250-ml. three-necked flask is equipped with a magnetic stirrer, a thermometer, a pressure-equalizing dropping funnel, and a reflux condenser with the entire system flushed with nitrogen. To the flask is added under an atmosphere of nitrogen 50.0 g. (0.454 mole) of phenylphosphine (Note 1), 50 ml. of acetonitrile, and 10 ml. of 10N potassium hydroxide (Note 2). An ice-water bath is prepared for immediate cooling of the reaction flask. To the reaction mixture is added dropwise 50.0 g. (0.94 mole) of acrylonitrile (Note 3) with stirring and cooling over a period of 45–60 minutes. The rate of addition is controlled so that the temperature of the solution never exceeds 35° (Note 4). After the addition is complete, the solution is stirred at room temperature for an additional 2.5 hours. The reaction mixture is subsequently diluted with 100 ml. of ethanol and chilled to 0°. The product starts to crystallize, and the mixture is allowed to stand until crystallization is complete. The heavy slurry is filtered, and the crystalline product is washed on the filter with 200 ml. of cold ethanol to give, after drying at 60° (2 mm.), 74–84 g. (76–86%) of bis(2-cyanoethyl)phenylphosphine, m.p. 71–74° (Note 5). An additional 5–9 g. of product may be recovered from the combined washings and filtrate by concentration of the solution with subsequent chilling to bring the total yield to 79–91 g. (80–93%).

B. *Preparation of 4-Amino-1,2,5,6-tetrahydro-1-phenyl-phosphorin-3-carbonitrile*. To a nitrogen-flushed 1-l. three-necked flask equipped with a mechanical stirrer, a pressure-equalizing addition funnel, and a reflux condenser are added 25 g. (0.22 mole) of potassium *tert*-butoxide (Note 6) and 200 ml. of toluene (Note 7). The mixture is heated to reflux, and a solution of 43.2 g. (0.20 mole) of bis(2-cyanoethyl)phenylphosphine from Section A in 400 ml. of toluene is added dropwise with stirring over a period of 40–50 minutes (Note 8). After the addition is complete, the mixture is stirred and heated at reflux for an additional 3 hours. The mixture is subsequently cooled to room temperature, 250 ml. of water is added, and the resulting mixture is stirred for 30 minutes while the product crystallizes. The mixture is filtered, and the solid product is washed on the filter with two 50-ml. portions of cold ethanol to give, after drying at 78° (1 mm.), 36–38 g. (84–88%) of 4-amino-1,2,5,6-tetrahydro-1-phenylphosphorin-3-carbonitrile, m.p. 134.5–137° (Note 9). A small amount of product can be recovered from the filtrate (Note 10).

C. *Preparation of 1-Phenyl-4-phosphorinanone*. A solution of 35 g. (0.162 mole) of 4-amino-1,2,5,6-tetrahydro-1-phenyl-phosphorin-3-carbonitrile in 400 ml. of 6N hydrochloric acid is heated at a vigorous reflux under nitrogen for 30 hours in a 1-l. three-necked flask equipped with a mechanical stirrer and a reflux condenser (Note 11). The mixture is then cooled with an ice bath, and 300 ml. of cold 10N potassium hydroxide is added with stirring over a period of 10 minutes (Note 12). The resulting solution is stirred an additional 10 minutes (Note 13) and then extracted with 300 ml. of ether. The ether layer is separated and washed two times with 100-ml. portions of water. The ether layer is dried, and the solvent is removed by distillation using a rotary evaporator. The resulting oil crystallizes on standing to give 23–26 g. of crude product, m.p. 38–43°. Distillation through a short-path column yields 21.5–21.7 g. (68–69%) of pure 1-phenyl-4-phosphorinanone, m.p. 43.5–44°, b.p. 120–122° (0.02 mm.) (Note 14).

2. Notes

1. Phenylphosphine is available from Pressure Chemical Company and Strem Chemicals Inc. It is best stored in a dry box under nitrogen. The compound is extremely air sensitive and malodorous. The container should be handled in the hood while wearing rubber gloves. Satisfactory preparations of phenylphosphine have been described in the literature.[2]

2. Potassium hydroxide solution is prepared by adding 5.6 g. of potassium hydroxide to sufficient water to give a final volume of 10 ml.

3. Acrylonitrile available from Eastman Organic Chemicals, practical grade, is satisfactory.

4. The optimum reaction temperature is approximately 30°. A yellow product results at higher reaction temperatures, while lower reaction temperatures lead to an uncontrollable reaction resulting from the base-initiated polymerization of acrylonitrile.

5. This product is of satisfactory purity for the next step. If a purer product is desired, bis(2-cyanoethyl)phenylphosphine may be recrystallized from hot ethanol or distilled, b.p. 215–223° (0.2 mm.).[3] The infrared adsorption maxima (KBr wafer) occur at 3.24, 4.46, 6.75, 7.00, 7.50, 13.33, 13.96, and 14.40 μ. The proton n.m.r. spectrum (deuteriochloroform solution)

shows a complex multiplet centered at δ 7.2 (8H) and a complex multiplet centered at 7.6 (5H). The ^{31}P n.m.r. spectrum, determined in ethanol at 40.5 MHz., has a signal at $+21.4$ p.p.m. relative to 85% phosphoric acid.

6. Potassium *tert*-butoxide is available from MSA Research Corp.

7. Reagent grade toluene was dried by standing over sodium ribbon.

8. The reaction mixture becomes quite viscous when the addition is about two-thirds complete.

9. The product, as isolated, is pure enough for conversion to 1-phenyl-4-phosphorinanone. If a higher degree of purity is desired, the product may be recrystallized from ethanol-water or chromatographed on alumina. Infrared absorption maxima (KBr wafer) occur at 2.94, 2.99, 3.09, 3.48, 4.61, 6.09, 6.24, 7.15, 7.56, 8.42, 12.05, 12.80, 13.56, and 14.49 μ. The proton n.m.r. spectrum shows a 6-proton multiplet at δ 1.8–3.0, a broad 2-proton singlet at δ 5.15, and a 5-proton multiplet at δ 7.4–7.8 (deuteriochloroform solution containing a small amount of dimethyl sulfoxide-d_6).

10. A small amount of product can be recovered from the filtrate by extracting the water layer with chloroform.

11. A white precipitate forms in the reaction medium after approximately 6 hours of reaction time. This precipitate may be the hydrochloride salt of 1-phenyl-4-phosphorinanone and has a melting point in excess of 200°.

12. Rapid addition of base seems to result in higher yields than a more cautious addition, even though the temperature of the solution increases to about 40°. The solution must be strongly basic for efficient extraction.

13. Impure 1-phenyl-4-phosphorinanone may crystallize at this point. If crystallization occurs, the solid is recovered by filtration and washed thoroughly with two 20-ml. portions of water. The material is dried in a desiccator over phosphorous pentoxide to give a product of m.p. 42.5–44°. To obtain a product satisfactory for distillation, the checkers found it necessary to dissolve the material in ether and wash with water before distilling the product.

14. Gas chromatographic analysis of the product on a column containing 10% SE 30 on acid-washed Chromosorb U indicated one component, injected as a 20% solution in ethanol at 230° and helium flow of 15 ml. per minute to give a retention time of 760 seconds. The proton n.m.r. spectrum (deuteriochloroform solution) shows an 8-proton multiplet centered at δ 2.4 and a 5-proton multiplet centered at δ 7.4. Infrared absorption maxima (KBr wafer) occur at 3.25 (=CH), 3.36 and 3.42 (CH), 5.87 (C=O), 6.27 and 6.73 (aromatic C=C), 6.98 (P-phenyl), 13.27 and 14.27 μ (monosubstituted phenyl). The ^{31}P n.m.r. spectrum (CHCl$_3$ solvent) at 40.5 MHz. shows a signal at +39.3 p.p.m. relative to 85% phosphoric acid.

3. Discussion

This reaction sequence illustrates a broadly applicable synthetic route to a functionalized phosphorus heterocycle and has been utilized for the synthesis of the phenyl-,[4,5] ethyl-,[4] and methyl-4-phosphorinanones.[6]

The cyanoethylation of phenylphosphine has been carried out in the presence of a basic catalyst,[3] and at high temperature.[7,8] Mono(2-cyanoethyl)phenylphosphine has been reported as a contaminant but this difficulty has not been observed in the procedure reported herein.

The intermediates, bis(2-cyanoethyl)phenylphosphine and 4-amino-1,2,5,6-tetrahydro-1-phenylphosphorin-3-carbonitrile, are easily isolated and characterized and show little or no oxidation when exposed to the air. 1-Phenyl-4-phosphorinanone is a highly-crystalline material which is more sensitive to air oxidation than its two precursors. However, 1-phenyl-4-phosphorinanone may be stored in a well-capped bottle for several months without appreciable oxidation. It is best stored in a dark bottle in a dry box under nitrogen.

A further extension of this type of synthetic sequence is illustrated by the cyclization of 2-cyanoethyl(o-cyanophenyl)-phenylphosphine to the corresponding o-enaminenitrile followed by hydrolysis with acid to form 2,3-dihydro-1-phenyl-4(1H)-phosphinolinone.[9] The only other reported synthesis of this class of compounds involves the addition of phenylphosphine to substituted divinyl ketones.[10]

Phosphorinanones have been utilized as substrates for the preparation of alkenes,[11] amines,[12] indoles,[5,13] and in the synthesis of a series of secondary and tertiary alcohols via reduction,[10a] and by reaction with Grignard[6,11] and Reformatsky[11,14] reagents. Phosphorinanones have also been used as precursors to a series of 1,4-disubstituted phosphorins.[15] The use of 4-amino-1,2,5,6-tetrahydro-1-phenylphosphorin-3-carbonitrile for the direct formation of phosphorino-[4,3-*d*] pyrimidines has been reported.[16]

1. Department of Chemistry, Oklahoma State University, Stillwater, Oklahoma 74074. We gratefully acknowledge partial support of this work on a grant from the National Cancer Institute, CA 11967-08A1.
2. R. J. Horvat and A. Furst, *J. Amer. Chem. Soc.*, **74**, 562 (1952); L. D. Freedman and G. O. Doak, *J. Amer. Chem. Soc.*, **74**, 3414 (1952); F. G. Mann and I. T. Millar, *J. Chem. Soc.*, 3039 (1952).
3. M. M. Rauhut, I. Hechenbleikner, H. A. Currier, F. C. Schaefer, and V. P. Wystrach, *J. Amer. Chem. Soc.*, **81**, 1103 (1959).
4. R. P. Welcher, G. A. Johnson, and V. P. Wystrach, *J. Amer. Chem. Soc.*, **82**, 4437 (1960).
5. M. J. Gallagher and F. G. Mann, *J. Chem. Soc.*, 5110 (1962).
6. L. D. Quin and H. E. Shook, Jr., *Tetrahedron Lett.*, 2193 (1965).
7. B. A. Arbuzov, G. M. Vinokurova, and I. A. Perfil'eva, *Proc. Acad. Sci. USSR*, Chem. Sect., **127**, 657 (1959).
8. F. G. Mann and I. T. Millar, *J. Chem. Soc.*, 4453 (1952).
9. M. J. Gallagher, E. C. Kirby, and F. G. Mann, *J. Chem. Soc.*, 4846 (1963).
10. (a) R. P. Welcher and N. E. Day, *J. Org. Chem.*, **27**, 1824 (1962); (b) R. P. Welcher, U.S. Patent 3,105,096 (1963); [*C. A.*, **60**, 5553 (1964)].
11. H. E. Shook, Jr., and L. D. Quin, *J. Amer. Chem. Soc.*, **89**, 1841 (1967).
12. Don L. Morris and K. D. Berlin, *Phosphorus*, **2**, 305 (1972).
13. M. J. Gallagher and F. G. Mann, *J. Chem. Soc.*, 4855 (1963).
14. L. D. Quin and D. A. Mathewes, *Chem. Ind. (London)*, 210 (1963).
15. G. Markl and H. Olbrich, *Angew. Chem.*, **78**, 598 (1966); *Angew. Chem. Int. Ed. Engl.*, **5**, 589 (1966).
16. Theodore E. Snider and K. D. Berlin, *Phosphorus*, **2**, 43 (1972).

PREPARATION OF α,β-UNSATURATED ALDEHYDES
via THE WITTIG REACTION:
CYCLOHEXYLIDENEACETALDEHYDE

Submitted by WATARU NAGATA,[1] TOSHIO WAKABAYASHI,[1] and YOSHIO HAYASE[2]

Checked by KYO ABE and S. MASAMUNE

1. Procedure

Into a 1-l. three-necked round-bottomed flask, fitted with a magnetic stirrer, dropping funnel, and nitrogen inlet, are placed 5.45 g. (0.12 mole) of sodium hydride (51% oil dispersion) (Note 1) and 30 ml. of dry tetrahydrofuran (Note 2). The system is flushed with nitrogen and a solution of 30.2 g. (0.12 mole) of diethyl 2-(cyclohexylamino)vinylphosphonate[3] in 90 ml. of dry tetrahydrofuran is added dropwise to the stirred mixture over a period of 15 minutes. During the addition the temperature is maintained at 0–5° with an ice bath. The mixture is further stirred for 15 minutes at 0–5° to ensure complete reaction. A solution of 10.3 g. (0.11 moles) of cyclohexanone (Note 3) in 70 ml. of dry tetrahydrofuran is added dropwise to the mixture over a period of 20 minutes so that the temperature does not exceed 5°. The mixture is stirred for an additional 90 minutes at 20–25° in a water bath. During the stirring a gummy precipitate of sodium diethyl phosphate is observed. The mixture is poured into 500 ml. of cold water and extracted with three 300-ml. portions of ether. The combined ether extracts are washed twice with 200 ml. of saturated aqueous salt solution, dried over anhydrous sodium sulfate, and the ether is distilled under reduced pressure (35 mm.) at 25–30°. The residue is dissolved in 300 ml. of benzene and transferred to a 3-l. three-necked round-bottomed flask equipped with a stirrer and a reflux condenser. To this solution is added a solution of 72 g. (0.57 mole) of oxalic acid dihydrate in 900 ml. of water (Note 4). The stirred mixture is refluxed for 2 hours under nitrogen, cooled, and transferred to a separatory funnel. The aqueous layer is extracted with two 300-ml. portions of ether. The combined organic extracts are washed with 200 ml. of water, then with 200 ml. of saturated aqueous salt solution, and dried over anhydrous sodium sulfate, and the solvent is distilled under reduced pressure (35 mm.) at 25–30°. The residue is transferred to a 30-ml. round-bottomed flask and distilled under reduced pressure through a 5-cm. Vigreux column to yield 10.8 g. (83%) (Note 5) of cyclohexylideneacetaldehyde, b.p. 78–84° (12 mm.), containing *ca.* 15% of isomeric cyclohexenylacetaldehyde (Note 6).

2. Notes

1. Sodium hydride (50–51% in mineral oil) was purchased from Metal Hydrides Inc. and used as 51%.

2. Reagent grade tetrahydrofuran was freshly distilled over sodium hydride before use. The checkers used lithium aluminum hydride to dry the solvent [see *Org. Syn.*, **46**, 105 (1966) for warning note].

3. Reagent grade cyclohexanone was redistilled.

4. When a more concentrated solution (72 g. of oxalic acid in 450 ml. of water) was used, the product contained larger amounts of the β,γ-isomer, cyclohexenylacetaldehyde. To

suppress this double bond isomerization, a 4–7% aqueous oxalic acid solution was used.

5. The yields were 80–85% in several runs.

6. The submitters found that analysis of the final product by gas chromatography indicated a 15% contaminant of the by-product, cyclohexenylacetaldehyde. The analysis was conducted on a column packed with 5% XE-60 on Chromosorb W at 120°. The retention times for cyclohexenylacetaldehyde and cyclohexylideneacetaldehyde were 1.3 and 3.3 minutes, respectively. The checkers found that the product contained 10–15% of cyclohexenylacetaldehyde by gas chromatographic analysis and 12–16% by n.m.r. spectral analysis (deuteriochloroform solution, tetramethylsilane reference), using the relative intensity of two signals (δ 9.53 and 9.97) due to the aldehydic protons of the two compounds. Reported physical constants are b.p. 58–62° (16 mm.) for cyclohexenylacetaldehyde and b.p. 80–85° (16 mm.) for cyclohexylideneacetaldehyde.[4]

3. Discussion

For conversion of ketones into α,β-unsaturated aldehydes containing two additional carbon atoms, several multistep processes *via* ethynyl or vinyl carbinol intermediates have been reported.[4–10] Although the overall yields obtained by these routes for the conversion of cyclohexanone into cyclohexylideneacetaldehyde have never exceeded 50%, they were the only useful methods for this type of conversion until the excellent process of Wittig[11] appeared. This process consists of normal aldol condensations of ketones with the lithium salt of ethylidenecyclohexylamine and subsequent dehydration and hydrolysis.

The present procedure also illustrates an excellent general method for the conversion of ketones and aldehydes[12] into the corresponding α,β-unsaturated aldehydes using diethyl 2-(cyclohexylamino)vinylphosphonate.[3] The yield of product is usually high, and the reaction proceeds stereoselectively to afford only the *trans* isomer. In the reaction of 3-ketosteroids with this reagent, no β,γ-isomers were formed.[12] Recently

Meyers and co-workers[13] reported a new method for the synthesis of α,β-unsaturated aldehydes.

1. Shionogi Research Laboratory, Shionogi & Company, Ltd., Fukushima-ku, Osaka, 553 Japan.
2. Faculty of Science, Osaka City University, Sumiyoshi-ku, Osaka, 558 Japan.
3. W. Nagata, T. Wakabayashi, and Y. Hayase, *Org. Syn.*, **53**, 44 (1973).
4. J. B. Aldersley and G. N. Burkhardt, *J. Chem. Soc.*, 545 (1938).
5. E. A. Braude and O. II. Wheeler, *J. Chem. Soc.*, 320 (1955).
6. M. Julia and J.-M. Surzur, *Bull. Soc. Chim. Fr.*, 1615 (1956).
7. H. G. Viehe, *Chem. Ber.*, **92**, 1270 (1959).
8. G. Saucy, R. Marbet, H. Lindlar, and O. Isler, *Helv. Chim. Acta*, **42**, 1945 (1959), cf. V. T. Ramakrishnan, K. V. Narayanan, and S. Swaminathan, *Chem. Ind. (London)*, 2082 (1967).
9. M. C. Chaco and B. H. Iyer, *J. Org. Chem.*, **25**, 186 (1960).
10. A. Marcou and H. Normant, *Bull. Soc. Chim. Fr.*, 1400 (1966).
11. (a) G. Wittig, H.-D. Frommeld, and P. Suchanek, *Angew. Chem.*, **75**, 978 (1963); (b) G. Wittig and H. Reiff, *Angew. Chem. Int. Ed. Engl.*, **7**, 7 (1968).
12. W. Nagata and Y. Hayase, *Tetrahedron Lett.*, 4359 (1968); *J. Chem. Soc. C*, 460 (1969).
13. A. I. Meyers, A. Nabeya, H. W. Adickes, J. M. Fitzpatrick, G. R. Malone, and I. R. Politzer, *J. Amer. Chem. Soc.*, **91**, 764 (1969).

REDUCTION OF ALKYL HALIDES AND TOSYLATES WITH SODIUM CYANOBOROHYDRIDE IN HEXAMETHYLPHOSPHORAMIDE (HMPA):
A. 1-IODODECANE TO *n*-DECANE
B. 1-DODECYL TOSYLATE TO *n*-DODECANE

$$CH_3(CH_2)_8CH_2I \xrightarrow[\text{HMPA}]{\text{NaBH}_3\text{CN}} CH_3(CH_2)_8CH_3$$

$$CH_3(CH_2)_{10}CH_2OTs \xrightarrow[\text{HMPA}]{\text{NaBH}_3\text{CN}} CH_3(CH_2)_{10}CH_3$$

Submitted by ROBERT O. HUTCHINS,[1] CYNTHIA A. MILEWSKI, and BRUCE E. MARYANOFF
Checked by RONALD I. TRUST and ROBERT E. IRELAND

1. Procedure

A. *Reduction of 1-Iododecane to n-Decane.*   In a dry 100-ml. three-necked flask equipped with a stirring bar, a thermometer, and a condenser protected with a drying tube are placed 25 ml.

of hexamethylphosphoramide (HMPA) (Note 1), 1-iododecane (2.7 g., 0.010 mole) (Note 2) and sodium cyanoborohydride (0.943 g., 0.015 mole) (Note 3). The solution is stirred at 70° for 2 hours, then diluted with 25 ml. of water and extracted with three 30-ml. portions of ether. The combined extracts are washed twice with water, dried over anhydrous magnesium sulfate, and then the solvent is removed by distillation on a steam bath through a 12-in. vacuum-jacketed Vigreux column (Note 4). The residue is distilled at reduced pressure in a short-path apparatus (*Caution! foaming*) to obtain 1.25–1.29 g. (88–90%) (Notes 5, 6) of *n*-decane, b.p. 68–70° (14 mm.); n^{20}D 1.4122, n^{26}D 1.4085 (lit.,[2] n^{25}D 1.4097) (Note 7).

B. *Reduction of 1-Dodecyl Tosylate to n-Dodecane*. In a dry 200-ml. three-necked flask equipped exactly as described in Section A are placed 50 ml. of hexamethylphosphoramide (HMPA), 1-dodecyl tosylate (6.80 g., 0.0201 mole) (Note 8), and sodium cyanoborohydride (5.02 g., 0.080 mole) (Note 3). The solution is stirred at 80° for 12 hours (Note 9), then diluted with 50 ml. of water, and extracted with three 60-ml. portions of hexane. The hexane solution is washed twice with water, dried over anhydrous magnesium sulfate, and then concentrated at reduced pressure with a rotary evaporator. Distillation of the residue through a short-path apparatus (Note 5) (*Caution! foaming*) affords 2.49–2.64 g. (73–78%) of *n*-dodecane, b.p. 79–81°; (3.75 mm.) n^{24}D 1.4217 (lit.,[3] n^{20}D 1.4219) (Note 7).

2. Notes

1. Commercial hexamethylphosphoramide was distilled from calcium hydride and stored over 13× molecular sieves (Linde).

2. Commercial 1-iododecane (Eastman Organic Chemicals) was filtered through activated charcoal and distilled before use.

3. Sodium cyanoborohydride was used as received from Aldrich Chemical Company, Inc. If other sensitive functional groups are present, it is advisable to purify the commercial reagent by the method of Purcell.[4]

4. If a hexane workup is used, and the solvent is removed with a rotary evaporator, considerable loss of product results

from codistillation with the hexane. This should not present a significant problem when higher boiling materials are produced.

5. The condenser was cooled with an ethylene glycol-water mixture at $-5°$, and the receiver was cooled to $-10°$ in an ice-salt bath.

6. Considerable mechanical loss was observed because of inability to distill the last portions of the product at 14 mm. To avoid this problem, the pressure was reduced near the end of the distillation to *ca.* 5 mm. This did not affect the purity of the product (Note 7).

7. Both products showed identical i.r. and n.m.r. spectra as those of authentic samples, and no side products were detected by g.l.p.c. or n.m.r.

8. Dodecyl tosylate was prepared from 1-dodecanol by the procedure of Marvel and Sekera.[5] Crystallization from a dried (magnesium sulfate) solution in light petroleum ether afforded white needles, m.p. 27.5–28.5°.

9. The large excess of sodium cyanoborohydride is recommended for the reduction of tosylates. Use of reduced molar excesses led to substantially lower yields. For example, a 3:1 cyanoborohydride to tosylate ratio afforded less than 60% yield of product at 80° for 5 hours, while a 1.5:1 excess gave only 52% yield at 70° for 8 hours.

3.   Discussion

These preparations that illustrate the use of sodium cyanoborohydride in hexamethylphosphoramide as an effective, selective, and convenient procedure for the reduction of alkyl halides and tosylates is essentially the same as previously described.[6] The very mild reducing ability of sodium cyanoborohydride makes the method particularly valuable when other functional groups are present in the molecule

$$\left(\text{i.e., } CO_2H, CO_2R, CN, NO_2, -\overset{\overset{O}{\|}}{C}N\!\!<, >\!\!C\!\!=\!\!O, -\overset{|}{C}\overset{O}{\diagdown\diagup}\overset{|}{C}-\right)^{.6}$$

In addition, alkene side-products are seldom encountered,

contrary to the situation with lithium aluminum hydride[7] or sodium borohydride in aqueous diglyme.[8] The combination of sodium borohydride in polar aprotic solvents is also effective for halide and tosylate removal,[9] although the possible selectivity is less.

1. Department of Chemistry, Drexel University, Philadelphia, Pennsylvania 19104.
2. A. F. Forziati, A. R. Glasgow, Jr., C. B. Willingham, and F. D. Rossini, *J. Res. Nat. Bur. Stand.*, **36**, 129 (1946) [*C.A.* **40**, 4341[9] (1946)].
3. A. F. Shepard, A. L. Henne, and T. Midgley, Jr., *J. Amer. Chem. Soc.*, **53**, 1948 (1931).
4. R. C. Wade, E. A. Sullivan, J. R. Bershied, Jr., and K. F. Purcell, *Inorg. Chem.*, **9**, 2146 (1970).
5. C. S. Marvel and V. C. Sekera, *Org. Syn.*, Coll. Vol. **3**, 366 (1955).
6. R. O. Hutchins, B. E. Maryanoff, and C. A. Milewski, *J. Chem. Soc.* (*D*), 1097 (1971).
7. N. G. Gaylord, "Reductions with Complex Metal Hydrides," Interscience Publishers, Inc., New York, 1956.
8. H. M. Bell and H. C. Brown, *J. Amer. Chem. Soc.*, **88**, 1473 (1966).
9. R. O. Hutchins, D. Hoke, J. Keogh, and D. Koharski, *Tetrahedron Lett.*, 3495 (1969).

SELECTIVE α-BROMINATION OF AN ARALKYL KETONE WITH PHENYLTRIMETHYLAMMONIUM TRIBROMIDE: 2-BROMOACETYL-6-METHOXYNAPHTHALENE AND 2,2-DIBROMOACETYL-6-METHOXYNAPHTHALENE

(2-Bromo-6′-methoxy-2′-acetonaphthone and 2,2-dibromo-6′-methoxy-2′-acetonaphthone)

$$C_6H_5N(CH_3)_2 \xrightarrow[\text{toluene}]{(CH_3)_2SO_4} C_6H_5\overset{\oplus}{N}(CH_3)_3\ CH_3SO_4^{\ominus} \xrightarrow[\text{HBr}]{Br_2}$$

$$C_6H_5\overset{\oplus}{N}(CH_3)_3Br_3^{\ominus}$$
(PTT)

Submitted by J. JACQUES and A. MARQUET[1]
Checked by DAVID WALBA and ROBERT E. IRELAND

1. Procedure

Caution! All operations should be carried out in a well-ventilated hood because dimethyl sulfate is highly toxic and the bromoketones are lachrymators and skin irritants.

A. *Phenyltrimethylammonium Sulfomethylate.* A solution of 24.8 g. (26 ml., 0.2 mole) of freshly distilled dimethylaniline (Note 1) in 100 ml. of toluene (Note 2) is prepared in a 250-ml. Erlenmeyer flask which is equipped with a thermometer and a magnetic stirrer. The mixture is stirred and heated to about 40°. The heating is stopped and 19 ml. (0.2 mole) of distilled dimethyl sulfate (Note 3) is added through an addition funnel in about 20 minutes. After a few minutes, the colorless sulfomethylate starts to crystallize. The temperature which varies very little during the addition, rises slowly for one hour thereafter and reaches about 50°. The reaction is allowed to proceed at ambient temperature for 1.5 hours after the addition is complete and then it is heated on a steam bath for one hour. After cooling, the phenyltrimethylammonium sulfomethylate is filtered, washed with 20 ml. of dry toluene and dried under vacuum; yield 44–46.5 g. (89–94%) (Note 4).

B. *Phenyltrimethylammonium Tribromide.* A solution of 10 g. (0.04 mole) of phenyltrimethylammonium sulfomethylate in 10 ml. of aqueous 48% hydrobromic acid diluted with 10 ml. of water is prepared in a 125-ml. Erlenmeyer flask equipped with a magnetic stirrer. Bromine (2.5 ml.) (Note 5) is added to the stirred solution from a dropping funnel in about 20 minutes. An orange-yellow precipitate forms immediately and the slurry is stirred at room temperature for 5–6 hours. The product, phenyltrimethylammonium tribromide (PTT) is filtered, washed with about 10 ml. of water and air-dried under an efficient hood. The crude PTT, *ca.* 15 g., is recrystallized from 25 ml. of acetic acid to give, after filtration and air-drying, 12.9–14.0 g. (86–93%) (Note 6) of orange crystals, m.p. 113–115°.

C. *2-Bromoacetyl-6-methoxynaphthalene.* To a solution of 1 g. (0.005 mole) of 2-acetyl-6-methoxynaphthalene[2] in 10 ml. of anhydrous tetrahydrofuran[3] (Note 7) contained in a 125-ml. Erlenmeyer flask is added 1.88 g. (0.005 mole) of PTT in small portions. About 10 minutes is required for this operation. A white precipitate forms immediately and the solution becomes pale yellow. After 20 minutes, 50 ml. of cold water is added, and the crystalline precipitate (Note 8) is filtered and washed with 10 ml. of water. The crude, white 2-bromoacetyl-6-methoxynaphthalene (*ca.* 1.3 g., m.p. 100–105°) is recrystallized from 32 ml. of cyclohexane to give 1.1 g. (79%) of crystalline product, m.p. 107–109° (lit. 107–108°)[4] (Notes 9 and 10).

D. *2,2-Dibromoacetyl-6-methoxynaphthalene.* To a solution of 1 g. (0.005 mole) of 2-acetyl-6-methoxynaphthalene[2] in 10 ml. of anhydrous tetrahydrofuran[3] (Note 7) contained in a 125 ml.

Erlenmeyer flask is added 3.76 g. (0.01 mole) of PTT in small portions over 10 minutes. A white precipitate forms and after one hour the solution is yellow. Cold water (50 ml.) is added and the crystalline product (Note 8) is filtered and washed with 10 ml. of water. The crude 2,2-dibromoacetyl-6-methoxynaphthalene (*ca.* 1.7 g., m.p. 110–117°) is recrystallized from 15 ml. of ethanol to give, after filtration and washing with 2 ml. of ethanol, 1.40–1.55 g. (78–87%) of slightly yellow product, m.p. 116.5–118° (lit. 118–119°)[5] (Notes 9 and 11).

2. Notes

1. Commercial dimethylaniline is redistilled, b.p. 78° (13 mm.).

2. Benzene can also be used, but toluene is preferable because of its lower toxicity.

3. Commercial dimethyl sulfate is distilled, b.p. 70° (13 mm.). A slight deficiency of dimethyl sulfate ensures the complete utilization of this toxic product.

4. This product is slightly hygroscopic, but no special precautions are required for handling.

5. Bromine (B & A, ACS Reagent Grade) was used without further purification.

6. The "active bromine" can be titrated according to the following procedure: about 300 mg. of PTT is dissolved in 50 ml. of acetic acid, 10 ml. of a 5% solution of KI in ethanol is added, and the liberated iodine is titrated with a $0.1N$ solution of $Na_2S_2O_3$. Percent "active bromine": calculated 42.5%; found 42.1–42.5%. The molecular weight of PTT is 375.96.

7. Tetrahydrofuran was purified and dried as previously described.[3] PTT is remarkably soluble in tetrahydrofuran (630 g. per l. at 20°). Under the same conditions, the solubility of the resulting phenyltrimethylammonium bromide is only 0.09 g. per l.

8. If the product precipitates as an oil, mere standing at room temperature may cause it to crystallize. If not, the addition of *ca.* 3 ml. of tetrahydrofuran, followed by swirling will usually induce crystallization.

9. *This product, like other bromoketones, can be very irritating to exposed skin.*

10. N.m.r. $(CDCl_3)$: δ 3.94 (s, 3, OCH_3), 4.54 (s, 2, $COCH_2Br$), 7.20 (m, 4, ArH), 7.90 (m, 1, ArH), 8.21 (m, 1, ArH).

11. N.m.r. $(CDCl_3)$: δ 3.93 (s, 3, OCH_3), 6.86 (s, 1, $COCHBr_2$), 7.20 (m, 4, ArH), 7.90 (m, 1, ArH), 8.50 (m, 1, ArH).

3. Discussion

Quaternary ammonium perhalogenides, being solid compounds, constitute halogen sources which are very convenient to handle. Of the different compounds studied and examples of their use which have been reported,[6] pyridinium hydrobromide perbromide[7] is the most popular. Phenyltrimethylammonium tribromide (PTT), the utility of which was recognized by Marquet and Jacques,[8] has the advantage of high stability and ease of preparation. The procedure herein described is a modification of that of Vorländer and Siebert.[9]

When dissolved in tetrahydrofuran, PTT (like pyridinium hydrobromide perbromide) is a source of $Br_3^{\ominus}$ ions, the properties of which are different from those of molecular bromine. In particular, it is much less electrophilic and less reactive toward aromatic rings and double bonds.[10] It is thus a selective brominating reagent for ketones[5] or ketals[5,11] when the molecule has double bonds or activated aromatic nuclei which would be attacked by bromine. The two examples of use of this reagent clearly differentiate between its reactivity from that of bromine. Reaction of bromine with 2-acetyl-6-methoxynaphthalene (in ether solution) gives a mixture, the main constituent of which is the product resulting only from ring bromination (2-acetyl-5-bromo-6-methoxynaphthalene).[4]

Many other examples have been described that illustrate the possibility of carrying out selective reactions with PTT which would be impossible with bromine; for example, 3β-acetoxy-20-oxo-21-bromopregna-5,16-diene can be obtained from 3β-acetoxy-20-ethylenedioxypregna-5,16-diene;[5] anisyl cyclohexyl ketone gives the α-bromoketone in very good yield with the aromatic ring remaining unattacked;[5] 5,7-

dimethoxyflavanone can be brominated in good yield at the position alpha to the keto group although the aromatic ring is activated by two methoxy groups.[12]

Anhydrous tetrahydrofuran contributes to the selectivity of the reagent because of the stability of $Br_3^{\ominus}$ in this solvent. Moreover, tetrahydrofuran acts as a buffer by reaction with the liberated hydrobromic acid which is why PTT in tetrahydrofuran can also be very useful if the molecule bears acid-sensitive functions. It must be emphasized that anhydrous tetrahydrofuran must be used because small amounts of water can greatly retard the rate of bromination of ketones with resulting decreased selectivity.

Recently a related brominating agent, pyrrolidone hydrobromide tribromide has been described. This reagent, also in tetrahydrofuran, gives ω-bromobenzalacetone from benzalacetone.[13]

1. Organic Chemistry of Hormones Laboratory, College of France, 75231 Paris 5, France.
2. L. Arsenijevic, Y. Arsenijevic, A. Horeau, and J. Jacques, *Org. Syn.*, this volume, p. 5.
3. *Org. Syn.*, **46**, 105 (1966).
4. A. Marquet, A. Horeau, J. Jacques, L. Novak, and M. Protiva, *Collect. Czech. Chem. Commun.*, **26**, 1475 (1961).
5. A. Marquet and J. Jacques, *Bull. Soc. Chim. Fr.*, 90 (1962).
6. A. Marquet, M. Dvolaitzky, H. B. Kagan, L. Mamlok, C. Ouannes, and J. Jacques, *Bull. Soc. Chim. Fr.*, 1822 (1961).
7. L. F. Fieser and M. Fieser, "Reagents for Organic Synthesis," John Wiley & Sons, Inc., New York, 1967, p. 967.
8. A. Marquet and J. Jacques, *Tetrahedron Lett.* (9) 24 (1959).
9. D. Vorländer and E. Siebert, *Ber.*, **52**, 283 (1919).
10. A. Marquet, J. Jacques, and B. Tchoubar, *Bull. Soc. Chim. Fr.*, 511 (1965).
11. W. J. Johnson, J. Dolf Bass, and K. L. Williamson, *Tetrahedron*, **19**, 861 (1963).
12. D. Brule and C. Mentzer, *C. R. Acad. Sci., Paris*, **250**, 365 (1960).
13. D. V. C. Awang and S. Wolfe, *Can. J. Chem.*, **47**, 706 (1969).

STEREOSELECTIVE SYNTHESIS OF TRISUBSTITUTED OLEFINS: ETHYL 4-METHYL-E-4,8-NONADIENOATE

(4,8-Nonadienoic acid, 4-methyl, *trans*, ethyl ester)

$$CH_2{=}CHCH_2CH_2Cl \xrightarrow[\text{ether}]{Mg} CH_2{=}CHCH_2CH_2MgCl$$

$$CH_2{=}C(CH_3)CHO \Big| \text{ ether}$$

$$\underset{\displaystyle CH_2{=}CHCH_2CH_2CH{-}C{=}CH_2}{\overset{\displaystyle OH \quad\; CH_3}{\qquad\qquad\qquad\;| \qquad\; |}}$$

$$CH_3C(OC_2H_5)_3 \Big/ \begin{array}{l} CH_3CH_2CO_2H, \\ 138\text{–}140° \end{array}$$

$$\left[\underset{\displaystyle O}{CH_2{=}CHCH_2CH_2{-}\overset{\overset{\textstyle H}{|}}{C}} \; \underset{\displaystyle CH_3}{\overset{\displaystyle OC_2H_5}{C}} \begin{array}{l} CH_2 \\ CH_2 \end{array} \right]$$

$$\underset{\displaystyle CH_2{=}CHCH_2CH_2}{\overset{\displaystyle H}{}} C{=}C \underset{\displaystyle CH_3}{\overset{\displaystyle CH_2CH_2CO_2C_2H_5}{}}$$

Submitted by RONALD I. TRUST and ROBERT E. IRELAND[1]
Checked by DAVID G. MELILLO and HERBERT O. HOUSE

1. Procedure

A. *2-Methyl-1,6-heptadien-3-ol.* To a dry three-necked 1-l. round-bottomed flask fitted with a mechanical stirrer, a reflux condenser with a nitrogen inlet tube, and a 125-ml. pressure-equalizing dropping funnel capped with a rubber septum is added 15.7 g. (0.646 g.-atom) of magnesium turnings. The flask is dried by heating with a flame while a stream of dry nitrogen is passed through the reaction vessel from the condenser and allowed to exit from a hypodermic needle inserted in the rubber septum. After drying, the hypodermic needle is removed and the flask is allowed to cool; a static nitrogen atmosphere is maintained in the reaction vessel for the remainder of the reaction. A small crystal of iodine and 450 ml. of anhydrous ether are added (Note 1). A solution of 49.4 g. (0.546 mole) of 4-chloro-1-butene (Note 2) in 50 ml. of anhydrous ether is then added from the dropping funnel, dropwise and with stirring. Sufficient external heat is applied to the reaction flask to keep the temperature of the reaction mixture at about 30°. After approximately 10–50% of the chloride solution has been added, a spontaneous reaction ensues as evidenced by the disappearance of the yellow iodine color, the appearance of a grey color in the reaction solution, and the commencement of gentle refluxing. The external heat is removed and the remainder of the chloride solution is added at a rate that maintains gentle refluxing. After the addition is complete, the reaction mixture is refluxed for 30 minutes and then a solution of 40.1 g. (0.572 mole) of methacrolein (Note 3) in 50 ml. of anhydrous ether is added, dropwise with stirring and refluxing during 45 minutes. Since the reaction with methacrolein is exothermic, the application of external heat may not be necessary to maintain refluxing during the addition of the aldehyde. During the addition the reaction mixture usually becomes cloudy. When the addition is complete, the reaction mixture is refluxed with stirring for 1.5 hours. The reaction mixture is cooled in an ice water bath and 250 ml. of aqueous 5% hydrochloric acid is added slowly and with stirring (Note 4). The organic layer is separated and the aqueous layer is extracted with four 200-ml. portions of ether. The combined organic solutions are then washed successively with 200 ml. of saturated aqueous sodium bicarbonate and with 200 ml. of saturated aqueous sodium chloride and dried over anhydrous sodium sulfate. The resulting ether solution is concentrated and the residual liquid is distilled under reduced pressure to separate 37.2–47.3 g. (54–69%) of 2-methyl-1,6-heptadien-3-ol as a colorless liquid, b.p. 85–88° (33 mm.), n^{25}D 1.4531–1.4535 (Note 5).

B. *Ethyl 4-Methyl-E-4,8-nonadienoate.*   A 500-ml. one-necked round-bottomed flask containing a magnetic stirring bar is fitted with a Claisen adapter, two thermometers, and a receiving flask as illustrated in Figure 1. To the flask is added 186 g.

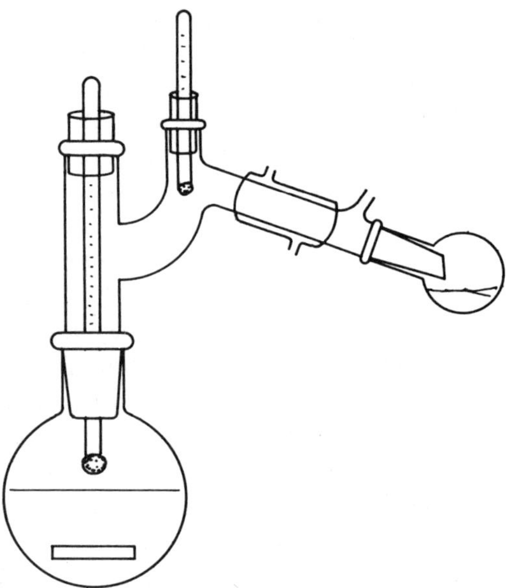

Figure 1.   Apparatus for the preparation of ethyl
4-methyl-E-4,8-nonadienoate.

(1.15 moles) of ethyl orthoacetate (Note 6), 25.2 g. (0.200 mole) of 2-methyl-1,6-heptadien-3-ol, and 0.70 g. (0.0094 mole) of propionic acid (Note 7). The mixture is heated with stirring to keep the temperature above the liquid at 138–142°. Heating is continued until ethanol no longer distils from the reaction flask (approximately one hour is required). The reaction mixture is then allowed to cool to room temperature and the excess ortho ester and propionic acid are removed by distillation under reduced pressure (approximately 50–60° at 20 mm.). The colorless to yellow residual liquid is then distilled under reduced pressure (0.25 mm., Note 8) to give 32.6–34.6 g. (83–88%) of ethyl 4-methyl-E-4,8-nonadienoate as a colorless liquid, b.p. 54–55° (0.25 mm.), n^{25}D 1.4504 (Note 9).

2. Notes

1. After drying, exposure of the reaction vessel and its contents to the atmosphere should be minimized. The iodine crystal should be added by lifting the dropping funnel and then replacing it quickly. The ether (anhydrous grade from Mallinckrodt Chemical Works) should be distilled from lithium aluminum hydride immediately before use and then transferred to the reaction vessel with a stainless steel cannula or a large hypodermic syringe inserted through the rubber septum.

2. 4-Chloro-1-butene is commercially available from Chemical Samples Company. The checkers employed this material without further purification. The submitters used material prepared from 3-buten-1-ol by a modified procedure of Roberts and Mazur.[2] Since material prepared according to the literature is invariably contaminated with thionyl chloride, which will interfere with formation of the Grignard reagent, the following modification is recommended. A two-necked 200-ml. round-bottomed flask is equipped with a magnetic stirring bar, a 60-ml. pressure-equalizing dropping funnel, and a reflux condenser fitted with a calcium chloride drying tube. The flask is charged with 49.8 g. (0.691 mole) of 3-buten-1-ol and 1.57 ml. of anhydrous pyridine (distilled from calcium hydride). With stirring and external cooling (ice water bath), 49 ml. (82 g., 0.69 mole) of thionyl chloride (Matheson Coleman and Bell commercial grade was used without further purification) is added dropwise over 3.5 hours. On completion of the addition, the mixture is heated under reflux for one hour. The external heating is then momentarily discontinued, and the condenser and dropping funnel are replaced by a stopper and short-path distilling head with receiver. Distillation of the mixture gives an opaque, colorless liquid (b.p. 68°). The crude product is washed with two 20-ml. portions of saturated aqueous sodium bicarbonate solution (frothing) and with 20 ml. of saturated

brine, and then dried over magnesium sulfate and filtered. The filtrate is distilled, and the 4-chloro-1-butene, collected as a colorless liquid boiling at 68–70°, amounts to 43.4 g. (67–69%).

3-Buten-1-ol. although commercially available from Aldrich Chemical Company, Inc., can be prepared economically and in large quantities by the addition of paraformaldehyde to allylmagnesium bromide[3] in ether according to procedures outlined for a similar synthesis.[4] In the present case, the submitters found it convenient to add the paraformaldehyde (Matheson Coleman and Bell commercial grade was dried overnight under reduced pressure and in the presence of phosphorus pentoxide) directly to the allylmagnesium bromide solution. After a reaction period of 6 hours at reflux, the previously described[4] isolation procedure gave 3-buten-1-ol in 56% yield.

3. Technical grade (90%) methacrolein (Aldrich Chemical Company, Inc.) was distilled (b.p. 67–69°) immediately before use.

4. Since the methacrolein is used in excess, frothing is no problem as there is no Grignard reagent remaining after the reaction is completed. Addition of 5% aqueous hydrochloric acid causes some coagulation of magnesium salts in the aqueous layer. These salts can be redissolved by addition of more aqueous 5% hydrochloric acid.

5. The product has the following spectral characteristics: i.r. (CCl_4), 3620 (free OH), 3480 (associated OH), 1645 (C=C), and 910 cm.$^{-1}$ (CH=CH_2); u.v. (95% C_2H_5OH) end absorption 210 mμ (ε 208); n.m.r. (CCl_4) δ 1.2–2.3 (m, 4, 2CH_2), 1.70 (s, 3, CH_3), 2.93 (broad, 1, OH), 4.00 (t, 1, $J = 6$ Hz., O—CH), 4.6–5.2 (m, 4, vinyl CH), and 5.5–6.1 (m, 1, vinyl CH); m/e (rel. int.), 111(28), 84(29), 83(21), 71(51), 71(100), 69(23), 67(28), 57(30), 55(51), 43(71), 41(49), and 39(37).

6. Ethyl orthacetate, available from Aldrich Chemical Company, Inc., was distilled before use. A large forerun was collected, consisting of hydrolysis products of the ortho ester. Material boiling at 135–142° is suitable for use in the reaction. It is convenient to transfer the material to the reaction flask

with a stainless steel cannula to avoid exposure of the ortho ester to atmospheric moisture. A fivefold excess of the ortho ester is needed, since the first step in the reaction is probably the reversible acid-catalyzed exchange of 2-methyl-1,6-hepta-dien-3-ol with ethanol.[5]

7. Practical grade propionic acid (Matheson Coleman and Bell) was distilled before use (b.p. 141°).

8. On two occasions, the submitters noticed a sublimable solid crystallizing in the distilling head just before the product began to distill. The distilling head was rinsed with ether, dried, and replaced, and the distillation was continued. The checkers observed the same phenomenon.

9. The product has the following spectral properties: i.r. (CCl_4) 1735 (ester $C{=}O$), 1645 ($C{=}C$), and 920 cm.$^{-1}$ ($CH{=}CH_2$); u.v. (95% C_2H_5OH) end absorption 210 mμ (ε 1960); n.m.r. ($CDCl_3$) δ 1.24 (t, 3, $J = 7$ Hz., OCH_2CH_3), 1.63 (broad, 3, $C{=}C{-}CH_3$), 1.9–2.2 (m, 4, $2CH_2$), 2.37 (broad, 4, $2CH_2$), 4.14 (q, 2, $J = 7$ Hz., OCH_2CH_3), 4.8–5.4 (m, 3, vinyl CH), and 5.5–6.2 (m, 1, vinyl CH); m/e (rel. int.), 196 (M^+, 4), 155(67), 151(30), 113(34), 109(100), 108(33), 85(47), 81(80), 67(74), 55(41), 53(31), 43(32), and 41(30). In C_6D_6 solution the allylic CH_3 signal of the major component present, the *trans*-isomer, is found at 1.50 and is accompanied by a minor peak at 1.61 attributable[6] to 3–4% of the *cis*-olefin in the product.

3. Discussion

The use of the Claisen rearrangement and several other methods for the stereoselective synthesis of trisubstituted olefins has been reviewed.[6] In allyl vinyl ethers of type A, the stereochemistry of the rearrangement is determined largely

by the steric requirements of R_1, which can be either axial or equatorial in the transition state.[7] When $R_3 = H$, the *trans/cis* ratio is approximately equal to the equatorial/axial equilibrium ratio of R_1-cyclohexane at the reaction temperature. When R_3 is larger than hydrogen, the steric effect is even greater, due to a potential 1,3-interaction which would develop in the transition state if R_1 were axial. No significant effect of R_2 on the *trans/cis* ratio has been observed.

The use of ethyl orthoacetate in the formation of vinyl ethers where $R_3 = OC_2H_5$ has been described.[5,8] The method described herein appears to be quite general, in that a variety of esters of type B ($R_3 = OC_2H_5$; $R_2 = CH_3$) may be prepared by merely varying the Grignard reagent used in preparing the starting allyl alcohol. The only limitations are the use of alcohols that are unsymmetrically bis-allylic from which mixtures of structural isomers may be obtained.

Stereoselectivity in the synthesis of trisubstituted olefins is necessary for the study of biosynthetic routes to polyisoprenoids, the nonenzymatic cyclization of polyolefinic substrates, and the study of insect hormones.

1. Department of Chemistry, California Institute of Technology, Pasadena, California 91109.
2. J. D. Roberts and R. H. Mazur, *J. Amer. Chem. Soc.*, **73**, 2509 (1951).
3. O. Grummitt, E. P. Budewitz, and C. C. Chudd, *Org. Syn.*, Coll. Vol. **4**, 748 (1963).
4. H. Gilman and W. E. Catlin, *Org. Syn.*, Coll. Vol. **1**, 188 (1944).
5. W. S. Johnson, L. Werthemann, W. R. Bartlett, T. J. Brocksom, T. Li, D. J. Faulkner, and M. R. Petersen, *J. Amer. Chem. Soc.*, **92**, 741 (1970).
6. D. J. Faulkner, *Synthesis*, 175 (1971).
7. D. J. Faulkner and M. R. Petersen, *Tetrahedron Lett.*, 3243 (1969).
8. W. S. Johnson, M. B. Gravestock, and B. E. McCarry, *J. Amer. Chem. Soc.*, **93**, 4332 (1971).

cis-α,β-UNSATURATED ACIDS: ISOCROTONIC ACID

[(*Z*)-Crotonic acid]

$$CH_3CH_2COCH_3 + 2\ Br_2 \longrightarrow CH_3CHBrCOCH_2Br + 2\ HBr$$

$$CH_3CHBrCOCH_2Br \xrightarrow[\text{2. HCl}]{\text{1. KHCO}_3} \underset{H}{\overset{CH_3}{\diagdown}}C=C\underset{H}{\overset{COOH}{\diagup}}$$

Submitted by C. RAPPE[1]
Checked by A. F. KLUGE and J. MEINWALD

1. Procedure

Caution! 1,3-Dibromo-2-butanone is a powerful lachrymator and a vesicant. This preparation should be carried out in a hood and contact of this compound with the skin should be avoided.

A. *1,3-Dibromo-2-butanone.* A mixture of 72.1 g. (90.0 ml., 1.0 mole) of 2-butanone and 100 ml. of precooled (5°) 48% hydrobromic acid is prepared in a 1-l. three-necked round-bottomed flask equipped with a dropping funnel, a condenser (Note 1), and a Teflon stirrer. The flask is immersed in ice water. When the temperature of the mixture reaches 5°, 319.6 g. (102 ml., 2.0 moles) of bromine is added dropwise at a rate such that the temperature does not rise above 10° and unreacted bromine does not accumulate (Note 2). After addition of the bromine is complete, 400 ml. of water is added, and the heavier organic layer is separated and immediately (Note 3) fractionated under reduced pressure (Note 4) through a 25-cm. Widmer column to give pure 1,3-dibromo-2-butanone, b.p. 91–94° (13 mm.), n^{25}D 1.5252 (Note 5). The yield is 115–134 g. (50–58%) (Note 6).

B. *Isocrotonic Acid.* To a solution of 100 g. (1.0 mole) of potassium bicarbonate (Note 7) in 1 l. of water contained in a 2-l. three-necked round-bottomed flask equipped with a condenser, a dropping funnel, and a Teflon stirrer, 46.0 g. (0.2 mole) of 1,3-dibromo-2-butanone is added over a 5-minute period

(Note 8). The mixture is stirred thoroughly, and after 2–3 hours (Note 9) when constant titration values against methyl orange are obtained, the solution is extracted with two 100-ml. portions of ether (Note 10) and acidified to pH 1–2 by the dropwise addition of dilute hydrochloric acid (Note 11). The aqueous solution is re-extracted with six 100-ml. portions of ether, and the ether phase is dried overnight in a refrigerator with magnesium sulfate.

The ethereal solution is filtered with suction, and the ether is removed under reduced pressure (water pump) on a rotary evaporator which is connected to a dry ice-acetone trap of 500-ml. capacity. A water bath maintained at 5–10° is used to facilitate the removal of the ether (Notes 12 and 13).

The yield of crude isocrotonic acid is 11.8–13.2 g. (69–77%). It is sufficiently pure for most purposes although n.m.r. analysis (Note 14) shows that the crude acid contains a small amount of the stable *trans*-isomer (Note 15). The crude product cannot be stored without isomerization.

For purification, 13.0 g. of the crude product is dissolved in 25 ml. of petroleum ether (b.p. 40–65°) at 5°. When left at $-15°$ for some days, crystals separate which are filtered at 5° to yield 9.3 g. of product, m.p. 12.5–14°, n^{25}D 1.4453. This sample can be stored in the dark at 30° for 3 weeks or at 5° for years with no detectable isomerization (Note 16).

2. Notes

1. The hydrogen bromide evolved from the condenser should be absorbed in a gas trap.

2. Accumulation of bromine results in an uncontrolled reaction and a decrease in the yield. This step requires 6–8 hours.

3. The crude product soon starts to decompose if it is not distilled immediately.

4. Since corrosive vapor is evolved, a water pump should be used.

5. The distilled product might be highly colored (violet, green, and blue), but this has no effect on its further use.

6. The dihaloketone purified in this way is stable for years when stored at 5°.

7. The yield is slightly lower when other bases such as sodium bicarbonate, sodium carbonate, and potassium carbonate are used.

8. The reaction is slightly exothermic.

9. Less time is required when stronger bases are used.

10. The nonacidic by-products are discarded.

11. Because of vigorous foaming, the addition must be made slowly and with care. The end-point can also be detected by a fading color of the reaction mixture. The checkers performed this acidification in the reaction flask with mechanical stirring which minimized the foaming.

12. The submitter used the following procedure for removal of ether. A 250-ml. two-necked round-bottomed flask, equipped with a dropping funnel, is arranged for distillation under reduced pressure (water pump). The ethereal solution is added dropwise (Note 13) and, when all the solution is added and the pressure has dropped to 10 mm., the last traces of ether are removed with an oil pump (0.4 mm.) for a period of 30 minutes.

13. This is to avoid isomerization which is easily initiated at elevated temperature.

14. N.m.r. spectroscopy is an excellent tool for distinguishing between the isomers.[2]

15. The checkers detected the presence of approximately 10% of the *trans*-acid by n.m.r. analysis.

16. The crude acid could be distilled in 5–10 ml. portions at 1 mm. without isomerization (b.p. 36°), but these samples were found to be more sensitive to isomerization.

3. Discussion

Isocrotonic acid can be prepared by the stereospecific *cis*-hydrogenation of tetrolic acid[3] or, mixed with the *trans*-isomer, by reduction of 3-chloro-*cis*-crotonic acid with sodium amalgam.[4] The *cis*-acid can also be prepared in small amounts by isomerization of the *trans*-acid.[5] The method herein described is much less laborious than the older procedures.[2]

TABLE I

PREPARATION OF BROMOKETONES AND cis-α,β-UNSATURATED ACIDS FROM METHYL KETONES

$$RCH_2COCH_3 \xrightarrow{2\ Br_2} RCHBrCOCH_2Br \xrightarrow[\text{2. HCl}]{\text{1. KHCO}_3} \underset{H}{\overset{R}{\diagdown}}C=C\underset{H}{\overset{COOH}{\diagup}}$$

R	Bromoketone		cis-α,β-Unsaturated Acid			
	Yield, %	$n^{25}\text{D}$	Yield, %	B.p.,° mm.	$n^{25}\text{D}$	m.p.,°
C_2H_5	57	1.5176	68	39–41(0.4)	1.4473	−43
n-C_3H_7	51	1.5080	61	71–73(0.2)	1.4495	0–1
iso-C_3H_7	58	1.5099	85	59.5–60(0.2)	1.4420	15.5–17.5
n-C_4H_9	65	1.5043	64	69–70(0.4)	1.4515	−19
t-C_4H_9	49	1.5071	75	60–61(0.8)	1.4432	11–12
n-C_5H_{11}	40	1.5001	50	75–76(0.3)	1.4530	—
$(CH_3)_2CH(CH_2)_2$	35	1.4997	58	93–94(0.2)	1.4518	—
n-C_6H_{13}	41	1.4983	28	91–92(0.8)	1.4549	2–3

The reaction is an example of a stereospecific Favorskii rearrangement,[6] and seems to have general applicability for the preparation of *cis*-α,β-unsaturated acids.[7] Only a limited number of the higher homologues have previously been prepared by the more laborious stereospecific *cis*-hydrogenation of the corresponding acetylenic acid,[1,8,9] and moreover, in some cases, they seem to have been mixtures of the two geometric isomers. The rearrangements, starting with commercially available methyl ketones, yield the *cis*-isomer exclusively as determined by n.m.r. spectroscopy.[7] The higher homologues can be purified by distillation with minimal losses. Purified in this manner the samples can be stored at 5° for years without detectable isomerization. The yields and physical constants of the bromoketones and of the *cis*-α,β-unsaturated acids are given in Table I.

1. Institute of Chemistry, University of Umeå, S-901 87 Umeå, Sweden.
2. C. Rappe, *Acta. Chem. Scand.*, **17**, 2766 (1963).
3. M. Bourguel, *Bull. Soc. Chim. Fr.*, (4) **45**, 1067 (1929).
4. A. Michael and O. Schulthess, *J. Parkt. Chem.*, (2) **46**, 236 (1892).
5. R. Stoenmer and E. Robert, *Ber.*, **55**, 1030 (1922).
6. A. S. Kende, *Org. React.*, **11**, 261, (1960).
7. C. Rappe and R. Adeström, *Acta. Chem. Scand.*, **19**, 383 (1965).
8. H. Silwa and P. Maitte, *Bull Soc. Chim. Fr.*, 369 (1962).
9. J. A. Knight and J. H. Diamond, *J. Org. Chem.*, **24**, 400 (1959).

3-NITROPHTHALIC ACID[1]

(Phthalic acid, 3-nitro-)

HAZARD NOTE

It has been reported that this reaction can become extremely violent. Modified directions that minimize this hazard have been published.[2]

Reported by J. H. P. Tyman and A. A. Durrani, *Chem. Ind. (London)*, 664 (1972).

---

[1] P. J. Culhane and Gladys E. Woodward, *Org. Syn.*, Coll. Vol. 1, 408 (1941).

[2] R. K. Bentley, *Chem. Ind.* (*London*), 767 (1972).

AUTHOR INDEX

This index comprises the names of contributors to Volume 50, 51, 52, and 53. Numbers in boldface type denote the volume; numbers in ordinary type indicate the page of that volume.

SUBJECT INDEX

This index comprises material from Volumes 50, 51, 52, and 53 only; for subjects of previous volumes see Collective Volumes I through V.

Entries in small capital letters refer to the titles of individual preparations. Entries in ordinary type letters refer to principal products and major by-products, special reagents or intermediates (which may or may not be isolated), compounds mentioned in the text, Notes or Discussions as having been prepared by the method given, and apparatus described in detail or illustrated by a figure. Numbers in boldface type denote the volumes. Numbers in ordinary type indicate pages on which a compound or subject is mentioned in the indicated volume.

<u>Unchecked Procedures</u>

Received during the period July 1, 1972 - July 9, 1973

In accordance with a policy adopted by the Board of Editors beginning with Volume 50, which, as noted in the Editor's Preface, is intended to make procedures available more rapidly, procedures received by the Secretary during the year, whether subsequently accepted, modified, or rejected for publication in Organic Syntheses, will be made available for purchase at the price of $2 per procedure, prepaid upon request to the Secretary:

> Dr. Wayland E. Noland, Secretary
> Organic Syntheses
> School of Chemistry
> University of Minnesota
> Minneapolis, Minnesota 55455

Payment must accompany the order, and should be made payable to Organic Syntheses, Inc. (not to the Secretary). Purchase orders not accompanied by payment will not be accepted. Procedures may be ordered by number and/or title from the list which follows.

It should be emphasized that the procedures which are being made available are unedited and have been reproduced just as they are first received from the submitters   (Those few procedures received during the period covered which have already been checked and are included in this volume are not listed here in unchecked form). There is no assurance that the procedures listed here will ultimately check in the form available, and many of them have been or will be rejected for publication in Organic Syntheses either before or after the checking process. For this reason, Organic Syntheses can provide no assurance whatever that the procedures will work as described, and offers no comment as to what safety hazards may be involved. Consequently, more than usual caution should be employed in following the directions in the procedures.

Organic Syntheses welcomes, on a strictly voluntary basis, comments from persons who attempt to carry out the procedures. For this purpose, a Checker's Report form will be mailed out with each unchecked procedure ordered. Procedures which have been checked by or under the supervision of a member of the Board of Editors will continue to be published in the volumes of Organic Syntheses, as in the past. It is anticipated that many of the procedures in the list will be published (often in revised form) in Organic Syntheses in future volumes.

-Wayland E. Noland

1813 (Revised)   <u>Reductive Cleavage of Allylic Alcohols, Ethers,</u>
<u>or Acetates to Olefins: 3-Methyl-1-cyclohexene</u>

I. Elphimoff-Felkin and P. Sarda, Centre de la Recherche
Scientifique, Institut de Chimie des Substances Naturelles,
91-Gif-sur-Yvette, France

1822   <u>p-Bromination of Aromatic Amines.   4-Bromo-N,N-dimethyl-3-</u>
<u>trifluoromethylaniline</u>

G. J. Fox, G. Hallas, J. D. Hepworth, and K. N. Paskins, Faculty of
Science, Derby College of Art and Technology, Kedleston Road,
Derby DE3, 1GB, England, U.K.

1823   Dimerization and Trimerization of Norbornadiene by
Rhodium on Carbon

Joseph J. Mrowca, Ronald J. Roth, Nancy Acton, and Thomas J.
Katz, Department of Chemistry, Columbia University, New York,
N. Y.   10027

1824   Synthesis of Benzvalene

Thomas J. Katz, Ronald J. Roth, Nancy Acton, and E. Jang
Carnahan, Department of Chemistry, Columbia University, New York,
N. Y.   10027

1825   Cyclobutanone from Methylenecyclopropane _via_ Oxaspiro-

pentane

J. R. Salaun, J. Champion, and J. M. Conia, Laboratoire des
Carbocycles, Université de Paris-Sud, 91 Orsay, France

1826   Cyclobutane-1,2-dione from $\underline{n}$-Butyl Succinate _via_

1,2-Bis(trimethylsiloxy)cyclobutene

J. M. Denis, J. Champion and J. M. Conia, Laboratoire des
Carbocycles, Université de Paris-Sud, 91 Orsay, France

1827   <u>Cyclopropanecarboxaldehyde from 1,2-Bis(trimethyl-
        siloxy)cyclobutene <u>via</u> Cyclobutane-1,2-diol</u>

J. P. Barnier, J. Champion, and J. M. Conia, Laboratoire des
Carbocycles, Université de Paris-Sud, 91 Orsay, France

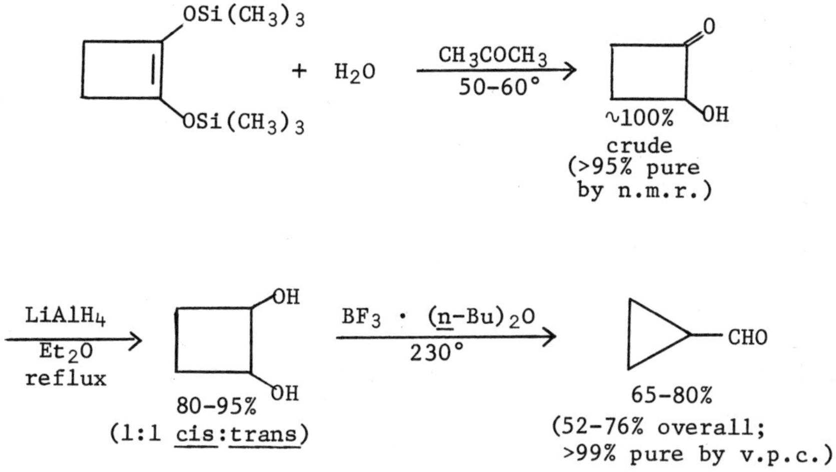

$OSi(CH_3)_3$ ... $+$ H_2O $\xrightarrow[50-60°]{CH_3COCH_3}$ ... O

$OSi(CH_3)_3$ ... ∼100% OH
crude
(>95% pure
by n.m.r.)

$\xrightarrow[\text{reflux}]{\substack{LiAlH_4 \\ Et_2O}}$ ... OH ... $\xrightarrow[230°]{BF_3 \cdot (\underline{n}\text{-Bu})_2O}$ ... CHO

80-95% OH
(1:1 <u>cis</u>:<u>trans</u>)

65-80%
(52-76% overall;
>99% pure by v.p.c.)

1828   <u>Methyl Groups by Reduction of Aromatic Carboxylic Acids
        with Trichlorosilane-Tri-<u>n</u>-Propylamine.   2-Methylbi-
        phenyl</u>

George S. Li, David F. Ehler, and R. A. Benkeser, Department of
Chemistry, Purdue University, Lafayette, Ind. 47907

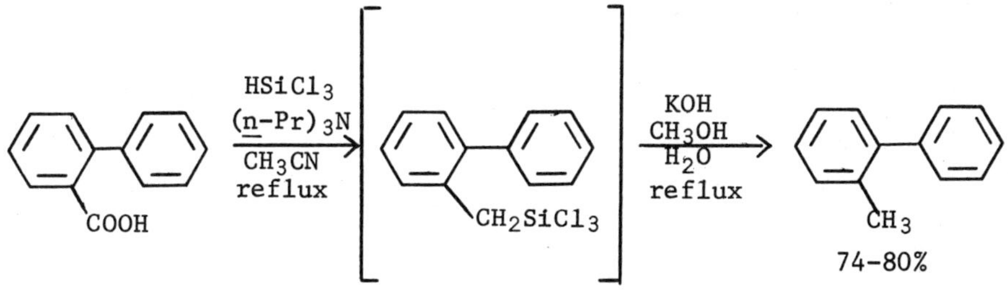

$\xrightarrow[\substack{CH_3CN \\ reflux}]{\substack{HSiCl_3 \\ (\underline{n}\text{-Pr})_3N}}$ ... CH_2SiCl_3 ... $\xrightarrow[\substack{H_2O \\ reflux}]{\substack{KOH \\ CH_3OH}}$ ... CH_3

COOH

74-80%

1829   <u>Substituted Hydrazobenzenes from Azobenzenes and
        Carbanions: N-(2,6-Dimethyl-4-pyridylmethyl)hydrazo-
        benzene</u>

Edwin M. Kaiser, Gregory J. Bartling, and Thomas M. Foy,
Department of Chemistry, University of Missouri-Columbia,
Columbia, Mo. 65201

$$\text{(2,4,6-trimethylpyridine)} + \varnothing N{=}N\varnothing \xrightarrow[\substack{\text{THF} \\ -78°}]{\substack{\text{n-BuLi} \\ \text{hexane}}} \xrightarrow{\substack{\text{NH}_4\text{Cl} \\ \text{H}_2\text{O}}} \text{(product)} \quad 65\text{--}77\%$$

1830   <u>tert-Butylacetylene</u>

J. R. Sowa, A. Gordinier, E. J. Lamby, D. A. Benko, and
E. G. Calamai, Department of Chemistry, Union College,
Schenectady, N. Y. 12308

$$\underline{\text{tert}}\text{-BuCl} + \text{CH}_2{=}\text{CHBr} \xrightarrow[-30°\ \text{to}\ -25°]{\text{AlCl}_3}$$

$$\underset{95\%}{\underline{\text{tert}}\text{-BuCH}_2\text{CHBrCl}} \xrightarrow[\substack{\text{DMSO} \\ 130-160°}]{\text{KOH}} \underset{\substack{73\% \\ (69\% \text{ overall})}}{\underline{\text{tert}}\text{-BuC}{\equiv}\text{CH}}$$

1833   <u>Ethyl 2-Oxocyclononanecarboxylate</u>

William L. Mock and Marvis E. Hartman, Department of Chemistry,
University of Illinois at Chicago Circle, Box 4348, Chicago,
Ill. 60680

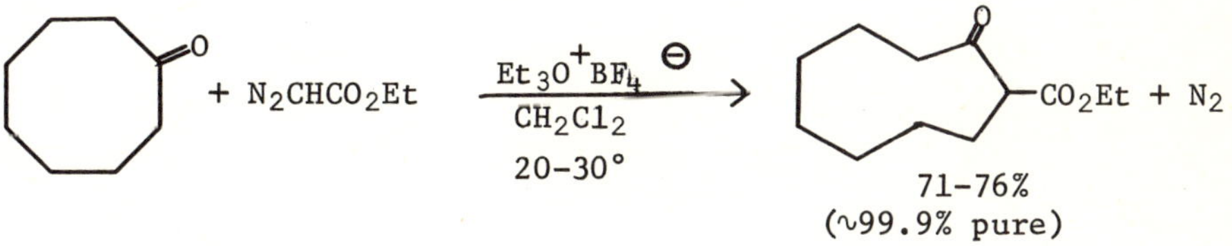

$$\text{cyclooctanone} + N_2CHCO_2Et \xrightarrow[\substack{CH_2Cl_2 \\ 20-30°}]{Et_3O^+BF_4^{\ominus}} \text{ethyl 2-oxocyclononanecarboxylate} + N_2$$

71-76%
(∼99.9% pure)

1834   <u>Indole-3-thiol</u>

R. L. N. Harris, Commonwealth Scientific and Industrial Research
Organization, P.O. Box 109, City, Canberra, A.C.T. 2601 Australia

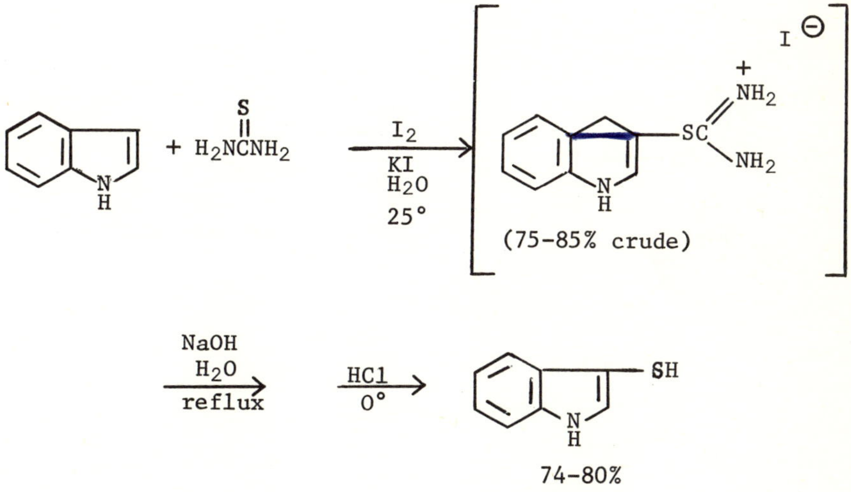

(75-85% crude)

NaOH
H$_2$O
reflux

HCl
0°

74-80%

1835   <u>Aldehydes from Olefins: Cyclohexanecarboxaldehyde</u>

P. Pino and C. Botteghi, Eidgenössische Technische Hochschule
Zürich, Technisch-Chemisches Laboratorium, Universitätstrasse 6,
8006 Zürich, Switzerland

$$\text{cyclohexene} + CO + H_2 \xrightarrow[\substack{150 \text{ atm.} \\ 100°}]{Rh_2O_3} \text{cyclohexanecarboxaldehyde (CHO)}$$

95%

(≥98% pure by v.p.c.)

1837   <u>6,7-Dimethoxy-3-isochromanone</u>

J. Finkelstein and A. Brossi, Chemical Research Department,
Hoffmann LaRoche Inc., Nutley, N. J. 07110

$$(CH_3O)_2\text{-}C_6H_3\text{-}CH_2COOH + HCHO \xrightarrow[\substack{AcOH \\ H_2O \\ 90°}]{HCl} \text{6,7-dimethoxy-3-isochromanone}$$

81%

Robert A. Auerbach, David S. Crumrine, David L. Ellison, and
Herbert O. House, Department of Chemistry, Georgia Institute of
Technology, Atlanta, Ga. 30332

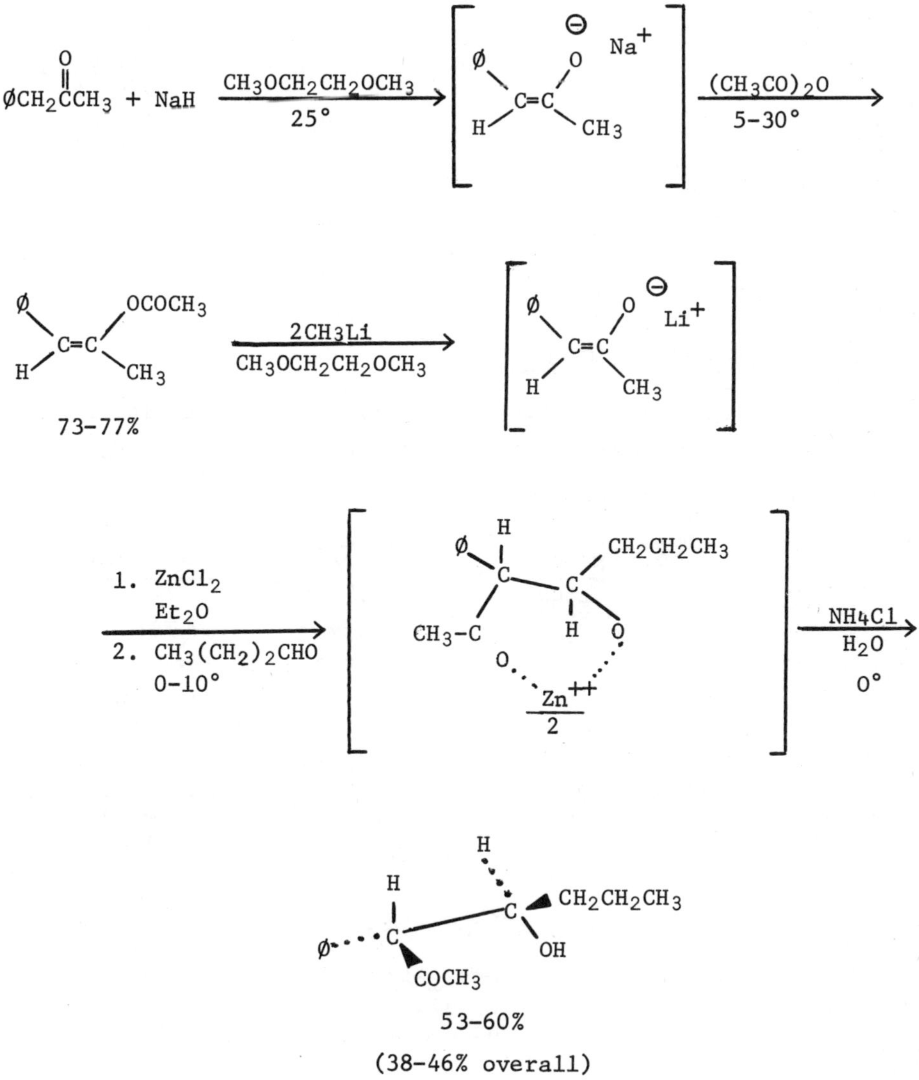

$$\phi CH_2\underset{\underset{O}{\parallel}}{C}CH_3 + NaH \xrightarrow[25°]{CH_3OCH_2CH_2OCH_3} \left[\ \phi,H\text{-}C=C\text{-}O^{\ominus},CH_3\ \right] Na^+ \xrightarrow[5-30°]{(CH_3CO)_2O}$$

enol acetate, 73–77%

$$\xrightarrow[CH_3OCH_2CH_2OCH_3]{2CH_3Li} \left[\ \phi,H\text{-}C=C\text{-}O^{\ominus},CH_3\ \right] Li^+$$

1. ZnCl$_2$, Et$_2$O
2. CH$_3$(CH$_2$)$_2$CHO, 0–10° $\longrightarrow$ [zinc chelate intermediate] $\xrightarrow[0°]{NH_4Cl,\ H_2O}$

53–60%

(38–46% overall)

1840   <u>Alkylations of Aldehydes via Reaction of the Magnesio-
Enamine Salt of an Aldehyde: 2,2-Dimethyl-3-phenyl-
propanal</u>

Gilbert Stork and Susan R. Dowd[*], (*) 234 Parkman Avenue,
Pittsburgh, Pa. 15213

$$(CH_3)_2CHCHO + H_2NC(CH_3)_3 \xrightarrow[25°]{N_2} (CH_3)_2CHCH=NC(CH_3)_3 \xrightarrow[\substack{2.\emptyset CH_2Cl \\ THF \\ reflux}]{\substack{1.EtMgBr \\ THF \\ reflux}}$$

50%

$$\xrightarrow[\substack{H_2O \\ reflux}]{HCl} \emptyset CH_2 \underset{\underset{CH_3}{|}}{\overset{\overset{CH_3}{|}}{C}} CHO$$

80%
(40% overall)

1841   <u>1,6-Methano[10]annulene</u>

E. Vogel, W. Klug, and A. Breuer, Institüt für Organische
Chemie, der Universität Köln, 47, Zulpicher Strasse 5, Cologne,
Federal Republic of Germany

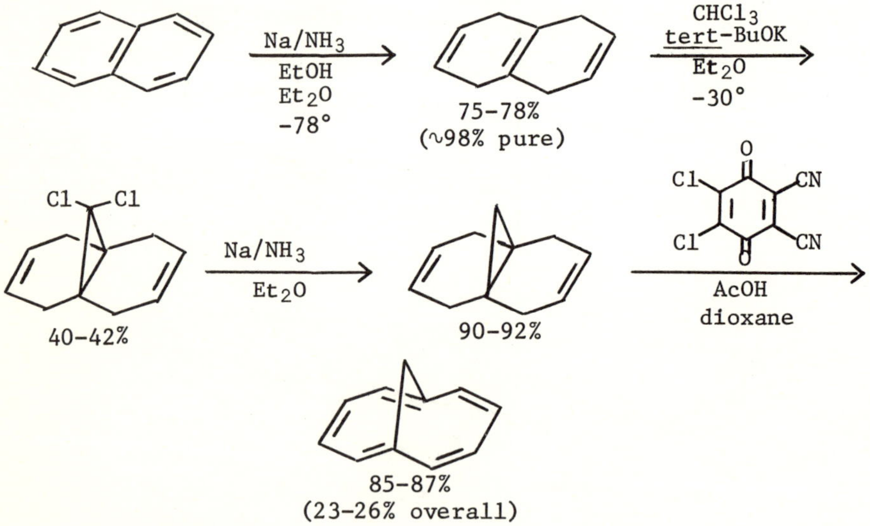

75–78%
(∼98% pure)

40–42%

90–92%

85–87%
(23–26% overall)

1842   <u>Olefins by Reaction of Lithium Dipropenylcuprates with</u>
<u>Alkyl Halides: trans-2-Undecene</u>

Gerard Linstrumelle, Jeanne K. Krieger, and George M. Whitesides,
Department of Chemistry, Massachusetts Institute of Technology,
Cambridge, Mass. 02139

$$\underset{H}{\overset{CH_3}{>}}C=C\underset{Cl}{\overset{H}{<}} \quad \xrightarrow[\substack{Et_2O \\ 13 \pm 3^\circ}]{Li(1\% \ Na)} \quad \left[\underset{H}{\overset{CH_3}{>}}C=C\underset{Li}{\overset{H}{<}}\right] \quad \xrightarrow[\substack{Et_2O \\ -78^\circ}]{CuI}$$

$$\left[\left(\underset{H}{\overset{CH_3}{>}}C=C\underset{}{\overset{H}{<}}\right)_2 CuLi\right] \quad \xrightarrow[\substack{[(CH_3)_2N]_3PO \\ Et_2O \\ -35^\circ}]{CH_3(CH_2)_6CH_2I} \quad \underset{H}{\overset{CH_3}{>}}C=C\underset{CH_2(CH_2)_6CH_3}{\overset{H}{<}}$$

90-93%
(96% <u>trans</u>, 4% <u>cis</u> by v.p.c.)

1843   Diethyl <u>tert</u>-Butyl(methyl)malonate

Charles M. Bartish and Charles S. Kraihanzel, Department of
Chemistry, Lehigh University, Bethlehem, Pa. 18105

$$(CH_3)_3CCH(COOEt)_2 \quad \xrightarrow[\substack{2. \ CH_3I}]{\substack{1. \ tert\text{-}BuOK \\ \varnothing CH_3}} \quad (CH_3)_3C\underset{CH_3}{\overset{|}{C}}(COOEt)_2$$

61-72%

1844   3-Methyl-2-norbornanone

D. L. Adams and W. R. Vaughan, Department of Chemistry,
University of Connecticut, Storrs, Conn.   06268

$$\text{(norbornanone)} \xrightarrow[\substack{HCl \\ H_2O \\ reflux}]{\substack{NH \\ CH_2O}} \text{(Mannich salt, } Cl^{\ominus}\text{)} \quad 35\text{-}59\% \xrightarrow{250\text{-}260°}$$

$$\text{(exomethylene ketone, } =CH_2\text{)} \quad 51\text{-}52\% \xrightarrow[EtOH]{\substack{H_2 \\ 5\% \; Pd\text{-}C}} \text{(3-methyl-2-norbornanone, } CH_3\text{)}$$

72-76%
(13-24% overall)

1845   Diphenylacetylene

J. R. Sowa and E. G. Calamai, Department of Chemistry, Union
College, Schenectady, N. Y.   12308

$$\underset{\substack{| \quad | \\ BrBr}}{\phi CHCH\phi} \xrightarrow[\substack{DMSO \\ 60°}]{CH_3ONa} \phi C\equiv C\phi \qquad 93\%$$

1846   Benzocyclopropene

W. E. Billups, A. J. Blakeney, and W. Y. Chow, Department of
Chemistry, Rice University, Houston, Tex.   77001

41-44%
(95% pure
by v.p.c.)

37%
(95% pure
by n.m.r.)

1847   3-Nitrophthalic Acid

R. K. Bentley, Research and Development Department, The Clayton
Aniline Co., Ltd., P.O. Box 2, Clayton, Manchester, England,
U.K.

25-26%
(containing <1% phthalic
acid or <2% 4-nitro-
phthalic acid by t.l.c.)

1848   Monoalkylation of α,β-Unsaturated Ketones <u>via</u> Metallo-

enamines. Preparation of 1,10-Dimethyl-Δ(1,9)-2-
octalone and 4-Methyl-Δ(4)-cholestenone

G. Stork and J. Benaim*,(*) Department of Chemistry, Centre
Universitaire de Toulon, Chateau Saint-Michel, 83-LaGarde,
France

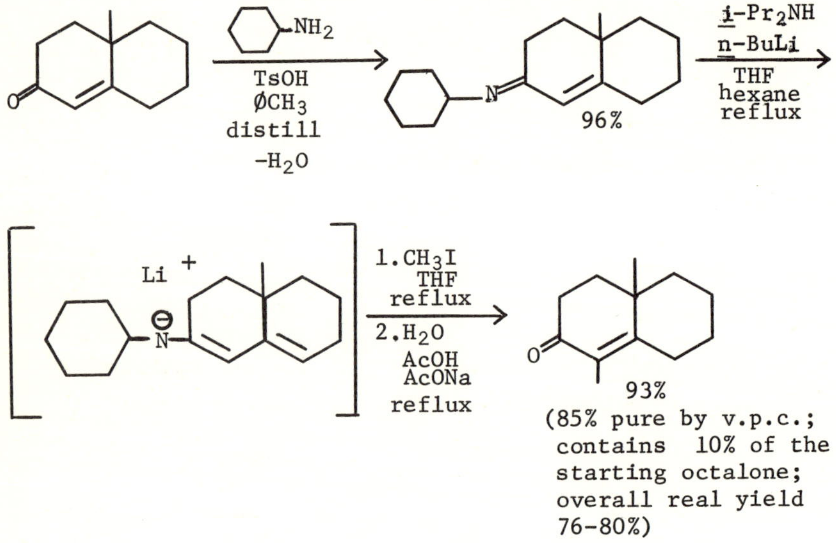

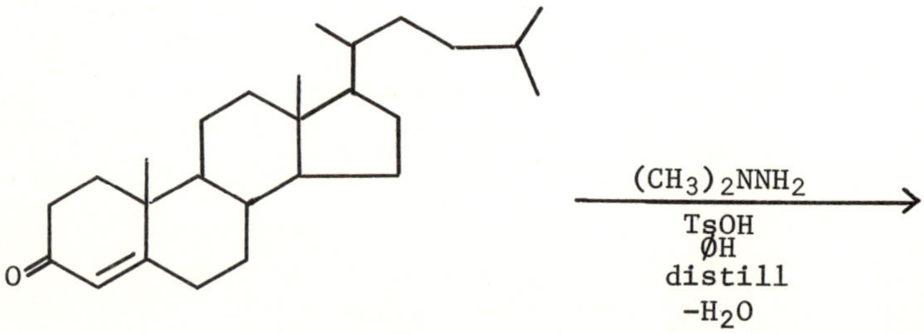

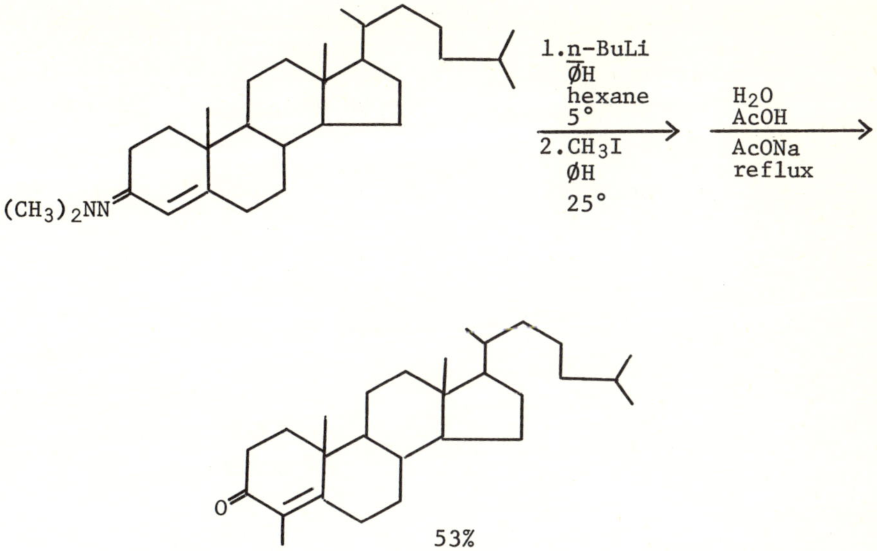

$(CH_3)_2NN$

1.n-BuLi
$\overline{\varnothing}$H
hexane
5°
2.CH$_3$I
$\varnothing$H
25°

H$_2$O
AcOH
AcONa
reflux

53%

1849   Endocyclic Enamine Synthesis: N-Methyl-2-phenyl-Δ^2-tetrahydropyridine

D. A. Evans and L. A. Domeier, Department of Chemistry, University of California, Los Angeles, Calif. 90024

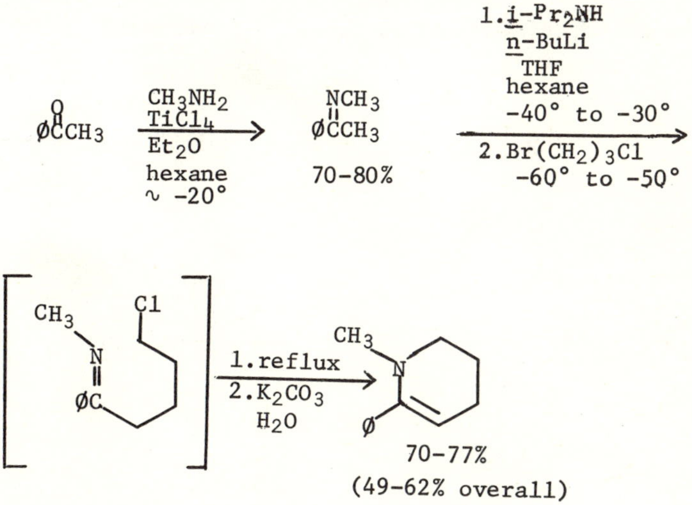

$\varnothing\overset{O}{\overset{\|}{C}}CH_3$

CH$_3$NH$_2$
TiCl$_4$
Et$_2$O
hexane
$\sim$ -20°

$\varnothing\overset{NCH_3}{\overset{\|}{C}}CH_3$

70-80%

1.$\underline{i}$-Pr$_2$NH
$\underline{n}$-BuLi
THF
hexane
-40° to -30°
2.Br(CH$_2$)$_3$Cl
-60° to -50°

1.reflux
2.K$_2$CO$_3$
H$_2$O

70-77%
(49-62% overall)

1850   <u>Diphenyldi(1,1,1,3,3,3-hexafluoro-2-phenyl-2-propoxy)-
sulfurane</u>

J. C. Martin, R. J. Arhart, J. A. Franz, E. F. Perozzi, and
L. J. Kaplan, Department of Chemistry, School of Chemical
Sciences, University of Illinois at Urbana-Champaign, Urbana,
Ill. 61801

$$\varnothing C(CF_3)_2OH \xrightarrow[\text{H}_2\text{O}]{\text{KOH}} \varnothing C(CF_3)_2O^{\ominus}\ K^+$$

98%

$$\xrightarrow[\text{CCl}_4]{\overset{\varnothing_2 S}{\text{Br}_2}}$$

65–75% conversion
(70–85% yield based
on unrecovered alcohol)

1851   <u>Di-<u>tert</u>-butyl Dicarbonate</u>

Barry M. Pope, Yutaka Yamamoto, and D. Stanley Tarbell,
Department of Chemistry, Vanderbilt University, Nashville, Tenn.
37203

$$(CH_3)_3COK \xrightarrow[\substack{\text{THF} \\ 0°}]{\text{CO}_2} [(CH_3)_3COCO_2K] \xrightarrow[\substack{\varnothing H \\ -15°}]{\text{COCl}_2}$$

$$(CH_3)_3COCO_2CO_2CO_2C(CH_3)_3 \xrightarrow[\substack{\text{CCl}_4 \\ -\text{CO}_2}]{} (CH_3)_3COCO_2CO_2C(CH_3)_3$$

55–64%   80%
(44–51% overall)

1852   <u>Oxidations with the Chromium Trioxide-Pyridine Complex</u>
      (Prepared <u>in situ</u>): Decyl Aldehyde

R. W. Ratcliffe, Research Laboratories, Merck & Company, Rahway,
N. J. 07065

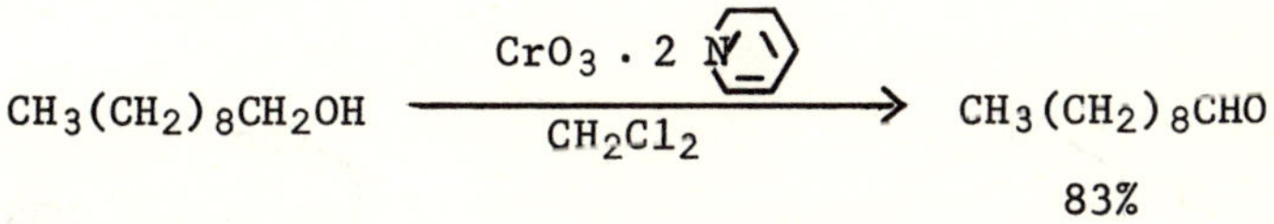

$$CH_3(CH_2)_8CH_2OH \xrightarrow[CH_2Cl_2]{CrO_3 \cdot 2 \text{ N}} CH_3(CH_2)_8CHO$$

83%

1853   <u>Cyclopropyldiphenylsulfonium Fluoroborate</u>

Mitchell J. Bogdanowicz and Barry M. Trost, Department of
Chemistry, University of Wisconsin, Madison, Wis. 53706

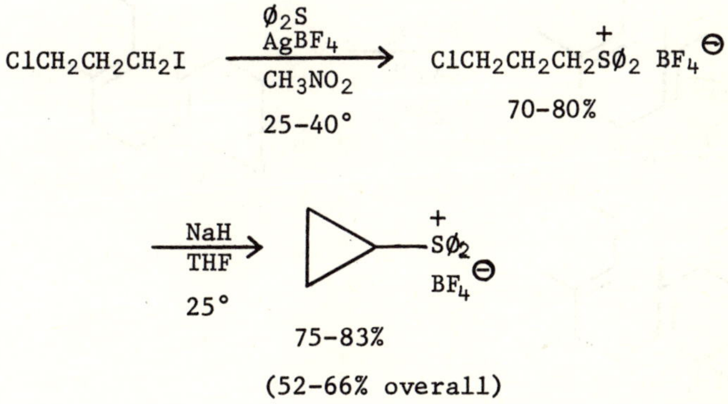

$$ClCH_2CH_2CH_2I \xrightarrow[\substack{CH_3NO_2 \\ 25-40°}]{\substack{\emptyset_2S \\ AgBF_4}} ClCH_2CH_2CH_2\overset{+}{S}\emptyset_2 \ BF_4^{\ominus}$$

70-80%

$$\xrightarrow[\substack{THF \\ 25°}]{NaH} \triangleright\hspace{-2pt}-\overset{+}{S}\emptyset_2 \ BF_4^{\ominus}$$

75-83%

(52-66% overall)

 <u>Selective Epoxidation of Terminal Double Bonds.</u>
<u>10,11-Epoxyfarnexyl Acetate</u>

R. P. Hanzlik, Department of Medicinal Chemistry, School of
Pharmacy, University of Kansas, Lawrence, Kan. 66044

OH

$\xrightarrow{\begin{array}{c}Ac_2O\\ \text{pyridine}\\ 25°\end{array}}$

OAc

$\sim100\%$

$\xrightarrow{\begin{array}{c}NBS\\ H_2O\\ \underline{tert-BuOH}\\ >12°\end{array}}$

OAc

Br —

HO   65%

$\xrightarrow{\begin{array}{c}K_2CO_3\\ CH_3OH\\ 25°\end{array}}$

OH

O

$\xrightarrow{\begin{array}{c}Ac_2O\\ \text{pyridine}\\ 25°\end{array}}$

OAc

O

60% overall

G. Nagendrappa, Engler Bunte Institüt der Universität Karlsruhe,
Bereich Petrochemie, Richard Willstätter Allee 5, Postfach 6380,
75 Karlsruhe 1, West Germany

$$\text{ClHC=CCl}_2 \quad 180\text{–}190° \qquad 22\text{–}24\%$$

$$\xrightarrow[\text{AcOEt}]{\overset{\text{H}_2}{5\%\ \text{Pd-C}}} \quad 84\text{–}91\%$$

$$\xrightarrow[\substack{\text{tert-BuOH} \\ 100\text{–}110°}]{\text{tert-BuOK}}$$

72–79%
(13–17% overall;
96–98% pure by v.p.c.)

Cycloaddition of <u>N</u>-Sulfonylamines to Alkenes. 6-Methyl-2-methoxy-1,4,3,5-oxathiadiazine 4,4-Dioxide. 2-Carbomethoxy-3,3-diphenyl-1,2-thiazetidine 1,1-Dioxide. N-Carbomethoxy-β,β-diphenylvinylsulfonamide

Edward M. Burgess and W. Michael Williams, Department of Chemistry, Georgia Institute of Technology, Atlanta, Ga.   30332

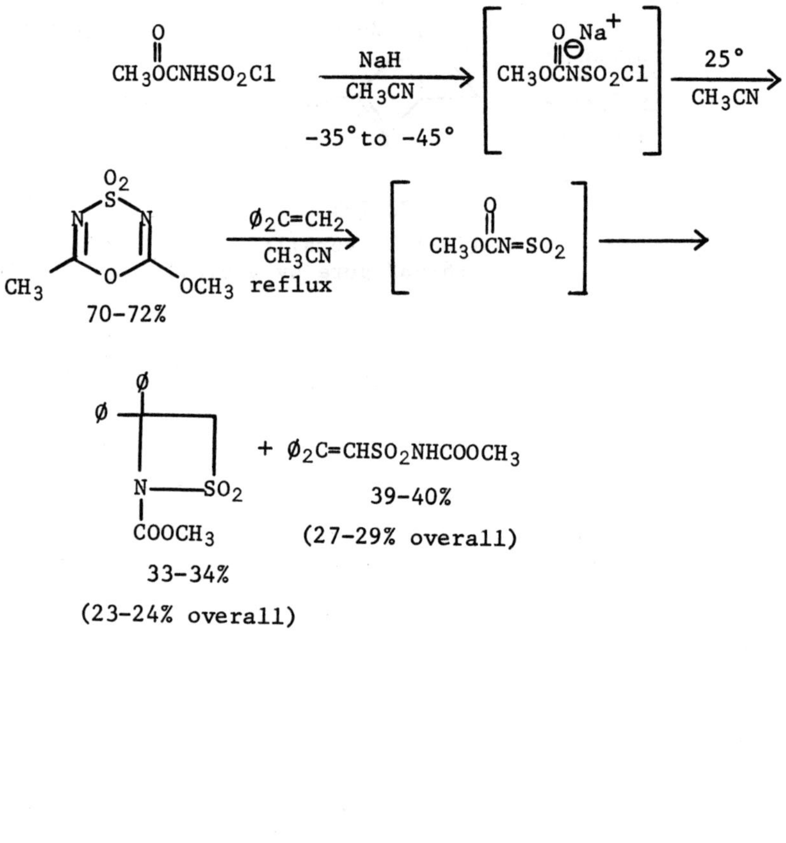

$$CH_3OCNHSO_2Cl \xrightarrow[\substack{CH_3CN \\ -35° \text{ to } -45°}]{NaH} \left[CH_3OCNSO_2Cl \right] \xrightarrow[CH_3CN]{25°}$$

$$\text{(6-methyl-2-methoxy-oxathiadiazine dioxide)} \xrightarrow[\substack{CH_3CN \\ reflux}]{\phi_2C=CH_2} \left[CH_3OCN=SO_2 \right] \longrightarrow$$

70-72%

(2-carbomethoxy-3,3-diphenyl-1,2-thiazetidine 1,1-dioxide) $+ \phi_2C=CHSO_2NHCOOCH_3$

33-34%

(23-24% overall)

39-40%

(27-29% overall)

1857   <u>Conversion of Primary Alcohols to Urethanes. Methyl</u>
         <u>N-Sulfonylurethane Triethylamine Complex</u>

Edward M. Burgess, Harold R. Penton, Jr., E. Alan Taylor, and
W. Michael Williams, Department of Chemistry, Georgia Institute
of Technology, Atlanta, Ga. 30332

$$OCNSO_2Cl \xrightarrow[\substack{\emptyset H \\ 25-30°}]{CH_3OH} CH_3O\overset{O}{\overset{\|}{C}}NHSO_2Cl \xrightarrow[\substack{\emptyset H \\ 25°}]{Et_3N}$$

$$88-92\%$$

$$CH_3O\overset{O}{\overset{\|}{C}}N\overset{\ominus}{S}O_2\overset{+}{N}Et_3 \xrightarrow[95\%]{CH_3(CH_2)_5OH}$$

$$84-87\%$$

$$CH_3O\overset{O}{\overset{\|}{C}}NH(CH_2)_5CH_3$$

$$55\%$$

$$(41-44\% \text{ overall})$$

1858   <u>1-Benzylindole</u>

Harry Heaney and Steven V. Ley, Organic Chemistry Laboratories,
University of Technology, Loughborough, Leicestershire LE11
3TU, England, U.K.

$$\text{(indole)} \xrightarrow[\substack{DMSO \\ 25°}]{\substack{ØCH_2Br \\ KOH}} \text{(1-benzylindole)}$$

$$92 - 97\%$$

1859   <u>Tricarbonyl(2,4-cyclohexadienone)iron and Tricarbonyl-
(2-methoxy-1,3-cyclohexadienylium)iron Fluorophosphate
from Anisole</u>

A. J. Birch and K. B. Chamberlain, Research School of Chemistry,
Australian National University, P.O. Box 4, Canberra, A.C.T.
2600 Australia

176

A. J. Birch and K. B. Chamberlain, Research School of Chemistry,
Australian National University, P.O. Box 4, Canberra, A.C.T.
2600 Australia

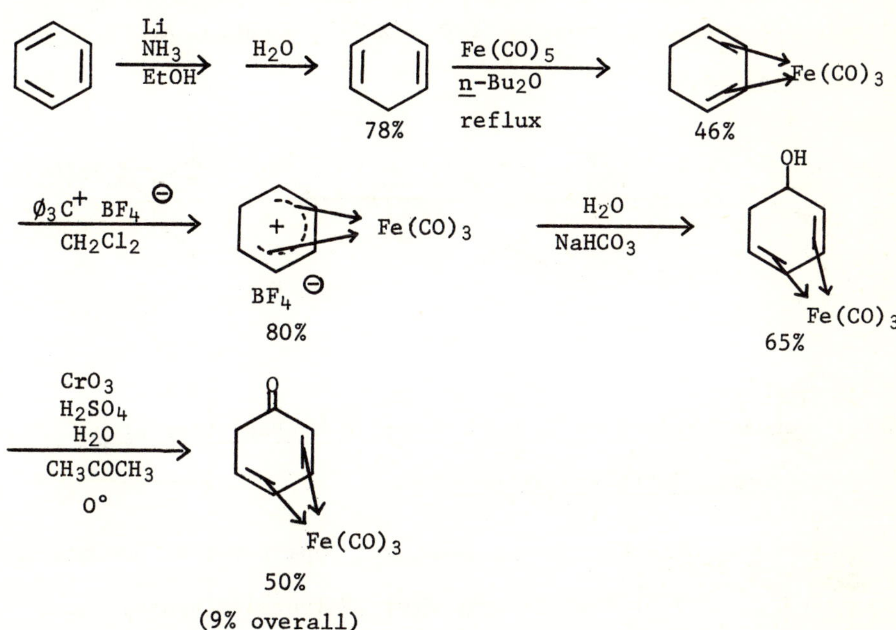

1861   <u>Tricarbonyl(1-ethoxycarbonylmethyl-1-hydroxy-2,4-cyclo-
        hexadiene)iron</u>

A. J. Birch and K. B. Chamberlain, Research School of Chemistry,
Australian National University, P.O. Box 4, Canberra, A.C.T.
2600 Australia

1862   Tricarbonyl[5-(2-hydroxy-4,4-dimethyl-6-oxo-1-cyclohexen-

1-yl)-2-methoxy-1,3-cyclohexadiene]iron

A. J. Birch and K. B. Chamberlain, Research School of Chemistry,

Australian National University, P.O. Box 4, Canberra, A.C.T.

2600 Australia

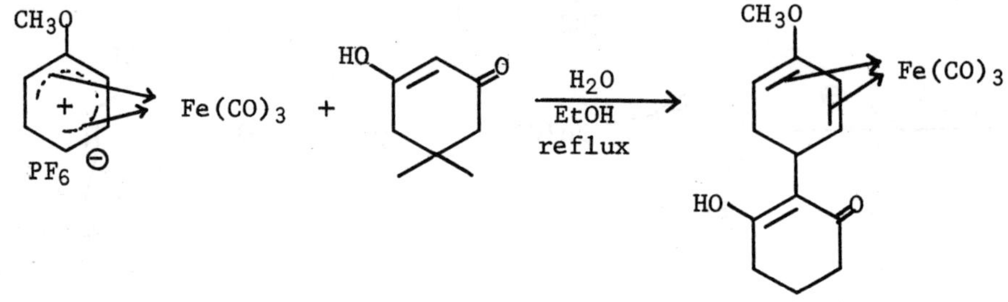

87%

1863   Benzaldehyde from Benzoyl Chloride by Reduction with

Disodium Tetracarbonylferrate

Y. Watanabe, T. Mitsudo, and Y. Takegami, Department of Hydrocarbon

Chemistry, Faculty of Engineering, Kyoto University, Kyoto, Japan

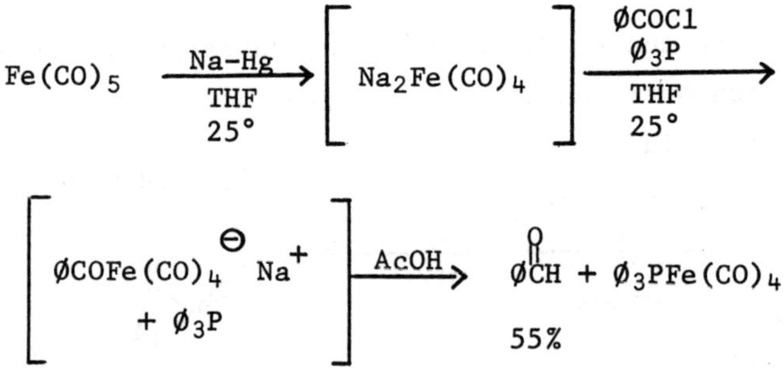

55%

     <u>trans-4-Hexen-1-ol by Sodium Ring-Scission of 3-Chloro-</u>

<u>2-methyltetrahydropyran</u>

Raymond Paul, Olivier Riobé, and Michel Maumy, Laboratoire de Chemie Organique, Ecole Superieure de Physique at de Chimie Industrielles, 10, rue Vauquelin, 75 - Paris v[e], France

$$\text{(tetrahydropyran)} \xrightarrow[\substack{Et_2O \\ -30° \text{ to } -10°}]{Cl_2} \left[\text{(dichloro intermediate)}\right] \xrightarrow[\substack{Et_2O \\ -15°}]{CH_3MgBr} \text{(3-chloro-2-methyltetrahydropyran)}$$

<u>cis</u> and <u>trans</u>

72%

$$\xrightarrow[\substack{2.\text{EtOH} \\ H_2O}]{\substack{1.\text{Na} \\ Et_2O \\ \text{reflux}}} \text{(trans-4-hexen-1-ol)}$$

89-94%
(64-68% overall;
pure by v.p.c.)

1865   <u>Ethyl 1-Cyano-2-methylcyclohexanecarboxylate (By Free
Radical Cyclization)</u>

Marc Julia and Michel Maumy, Laboratoire de Chimie Organique,
Ecole Superieure de Physique et de Chimie Industrielles, 10 rue
Vauquelin, 75 - Paris v^e, France

1866   <u>Geranyl Chloride</u>

Jose G. Calzada and John Hooz, Department of Chemistry, University
of Alberta, Edmonton 7, Alberta, Canada

1867   1-Ethoxy-1-butyne

Melvin S. Newman and Wayne M. Stalick, Department of Chemistry,
The Ohio State University, Columbus, Ohio 43210

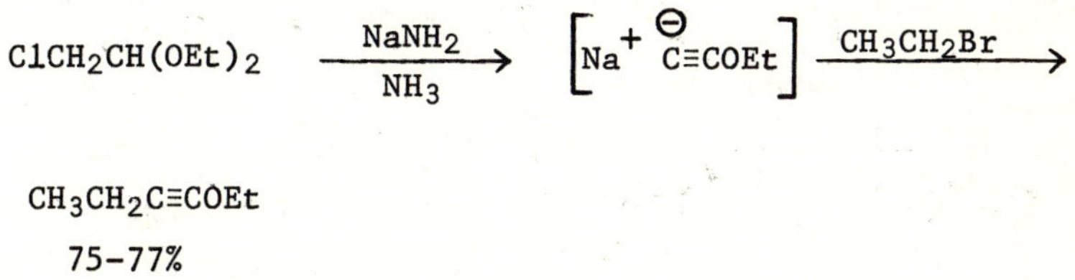

$$ClCH_2CH(OEt)_2 \xrightarrow[NH_3]{NaNH_2} \left[Na^+ \; {}^{\ominus}C\equiv COEt \right] \xrightarrow{CH_3CH_2Br}$$

$$CH_3CH_2C\equiv COEt$$
75–77%

1868   4-Chloro-4-hexen-1-yl Trifluoroacetate

P. E. Peterson and M. Dunham, Department of Chemistry, University
of South Carolina, Columbia, S. C.   28208

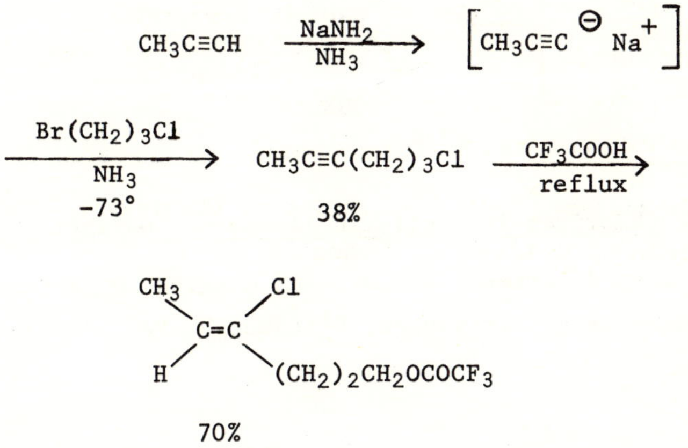

$$CH_3C\equiv CH \xrightarrow[NH_3]{NaNH_2} \left[CH_3C\equiv C^{\ominus} \; Na^+ \right]$$

$$\xrightarrow[\substack{NH_3 \\ -73°}]{Br(CH_2)_3Cl} \quad CH_3C\equiv C(CH_2)_3Cl \xrightarrow[reflux]{CF_3COOH}$$

$$38\%$$

$$\underset{H}{\overset{CH_3}{>}}C=C\underset{(CH_2)_2CH_2OCOCF_3}{\overset{Cl}{<}}$$

70%

Koichi Hirai and Yukichi Kishida, Central Research Laboratories, Sankyo Co., Ltd., 1-2-58, Hiromachi, Shinagawa-ku, Tokyo 140, Japan

(26% overall)

1870   Allylic Chlorides from Allylic Alcohols.   Geranyl Chloride

Gilbert Stork, Paul A. Grieco, and Michael Gregson, Department of Chemistry, University of Pittsburgh, Pittsburgh, Pa.   15213

1871   <u>Alkylation of Indolylsodium in Hexamethylphosphoramide</u>
        <u>(HMPA)</u>

George M. Rubottom and John C. Chabala, Department of Chemistry,
University of Puerto Rico, Rio Piedras, Puerto Rico 00931

$$\text{1. NaH} \quad [(CH_3)_2N]_3PO \quad 0\text{--}25° \qquad \text{2. } CH_3I \quad 0\text{--}25°$$

86-94%
(>99% pure by v.p.c.)

1872   <u>4-Ethoxy-3-hydroxybenzaldehyde</u>

Robert E. Ireland and David M. Walba, Department of Chemistry,
California Institute of Technology, Pasadena, Calif. 91109

$HOCH_2CH_2OH$, TsOH, $\varnothing H$, reflux, $-H_2O$

94%

1. $\varnothing_2PH$, n-BuLi, THF, hexane, 25°
2. HCl, H_2O

87-88%
(82-83% overall)

$+ \; \varnothing_2PCH_3$

1873 2-Phenoxymethyl-1,4-benzoquinone

Niels Jacobsen, Department of Organic Chemistry, Chemical Institute, University of Aarhus, 8000 Århus C, Denmark

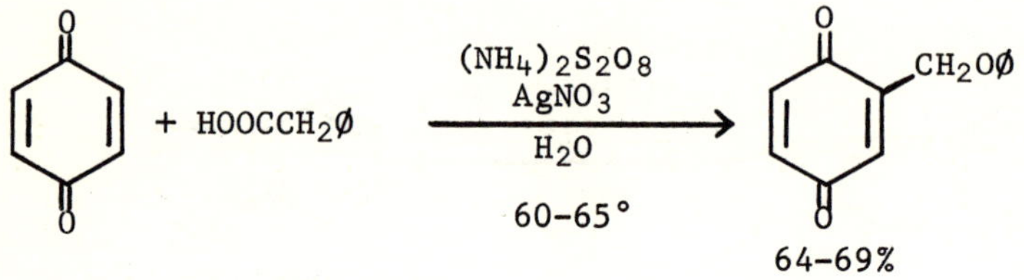
O
O
+ HOOCCH2Ø
(NH4)2S2O8
AgNO3
H2O
60-65°
O
CH2OØ
O
64-69%

1874 1,6-Oxido[10]annulene

E. Vogel, W. Klug, and A. Breuer, der Universität Köln, Institüt für Organische Chemie, 47, Zulpicher Strasse 5, Köln, Fed. Republic of Germany

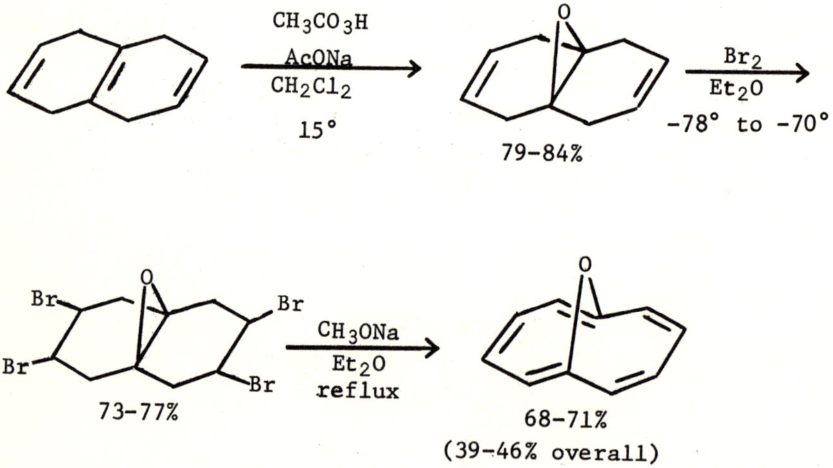
CH3CO3H
AcONa
CH2Cl2
15°
O
79-84%
Br2
Et2O
-78° to -70°
Br
Br
O
Br
Br
73-77%
CH3ONa
Et2O
reflux
O
68-71%
(39-46% overall)

1875   <u>2-Bromohexanoyl Chloride</u>

David N. Harpp, L. Q. Bao, Christopher Coyle, John G. Gleason, and
Sharon Horovitch, Department of Chemistry, McGill University
P.O. Box 6070, Montreal 101, Quebec, Canada

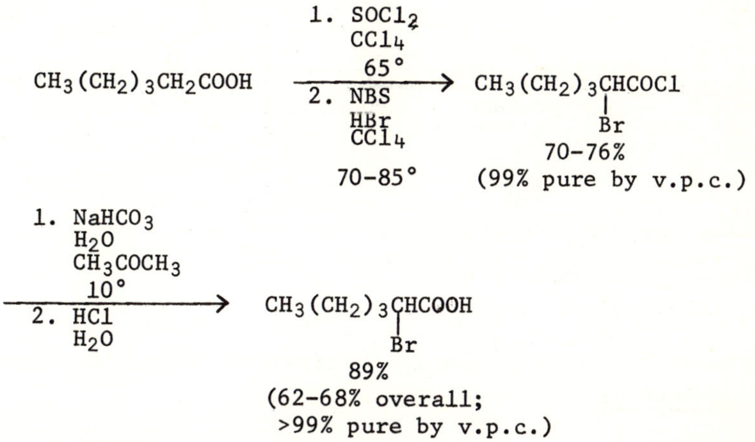

$$CH_3(CH_2)_3CH_2COOH \xrightarrow[\substack{2.\ NBS \\ HBr \\ CCl_4 \\ 70\text{-}85°}]{\substack{1.\ SOCl_2 \\ CCl_4 \\ 65°}} CH_3(CH_2)_3\underset{\underset{Br}{|}}{C}HCOCl$$

70-76%
(99% pure by v.p.c.)

$$\xrightarrow[\substack{2.\ HCl \\ H_2O}]{\substack{1.\ NaHCO_3 \\ H_2O \\ CH_3COCH_3 \\ 10°}} CH_3(CH_2)_3\underset{\underset{Br}{|}}{C}HCOOH$$

89%
(62-68% overall;
>99% pure by v.p.c.)

1876   <u>Stereospecific Reduction of Epoxides. A Route to Cationic</u>
       <u>Iron-Olefin Complexes</u>

M. Rosenblum and J. M. Tancrede, Department of Chemistry, Brandeis
University, Waltham, Mass. 02154

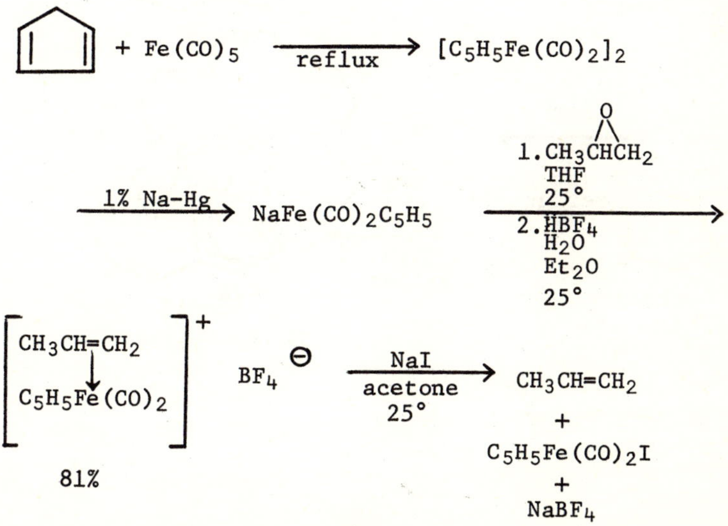

$$\text{(dicyclopentadiene)} + Fe(CO)_5 \xrightarrow{reflux} [C_5H_5Fe(CO)_2]_2$$

$$\xrightarrow{1\%\ Na\text{-}Hg} NaFe(CO)_2C_5H_5 \xrightarrow[\substack{2.\ HBF_4 \\ H_2O \\ Et_2O \\ 25°}]{\substack{1.\ CH_3\overset{O}{\overset{/\backslash}{CH}}CH_2 \\ THF \\ 25°}}$$

$$\left[\begin{array}{c} CH_3CH{=}CH_2 \\ \downarrow \\ C_5H_5Fe(CO)_2 \end{array} \right]^{+} BF_4^{\ominus} \xrightarrow[\substack{acetone \\ 25°}]{NaI} \begin{array}{c} CH_3CH{=}CH_2 \\ + \\ C_5H_5Fe(CO)_2I \\ + \\ NaBF_4 \end{array}$$

81%

1877   <u>The Addition of Organolithium Reagents to Allyl Alcohol:</u>
        <u>2-Methyl-1-hexanol</u>

J. K. Crandall and A. C. Rojas, Department of Chemistry, Indiana
University, Bloomington, Ind. 47401

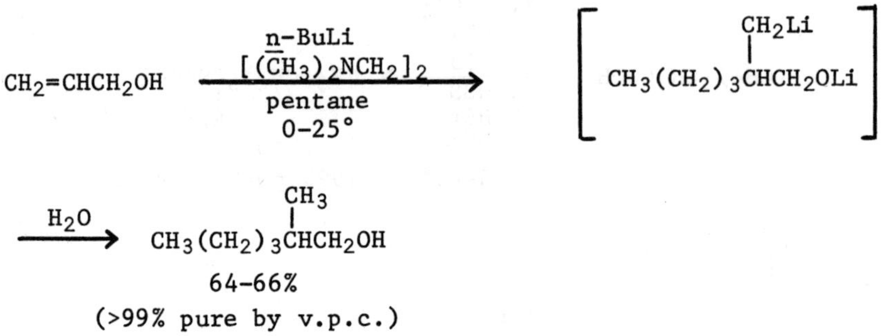

$$CH_2=CHCH_2OH \xrightarrow[\substack{\text{pentane} \\ 0\text{--}25°}]{\substack{n\text{--BuLi} \\ [(\overline{CH}_3)_2NCH_2]_2}} \left[CH_3(CH_2)_3\overset{\overset{\displaystyle CH_2Li}{|}}{C}HCH_2OLi \right]$$

$$\xrightarrow{H_2O} CH_3(CH_2)_3\overset{\overset{\displaystyle CH_3}{|}}{C}HCH_2OH$$

64–66%

(>99% pure by v.p.c.)

1878   <u>18-Crown-6</u>

George W. Gokel and Donald J. Cram, Department of Chemistry,
University of California at Los Angeles, Los Angeles, Calif. 90024

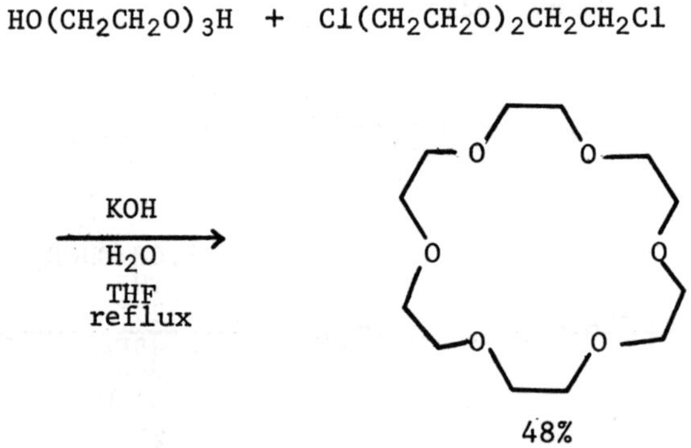

$$HO(CH_2CH_2O)_3H \;+\; Cl(CH_2CH_2O)_2CH_2CH_2Cl$$

$$\xrightarrow[\substack{H_2O \\ THF \\ reflux}]{KOH}$$

48%

1879   <u>Aldehydes from Acetals by Formolysis: Phenylpropargyl</u>
        <u>Aldehyde</u>

A. Gorgues, Laboratoire de Chimie Organique C, Université de
Rennes, Avenue du Général Leclerc, P.B. 25 A, 35031 Rennes-
Cedex-France

$$\emptyset C \equiv CCH(OEt)_2 \xrightarrow[100°]{HCOOH} \emptyset C \equiv CCHO + HCOOEt + H_2O$$
$$86\%$$

1880   <u>Substituted Hydrazobenzenes from Azobenzenes and Carbanions:</u>
        <u>N-Diphenylmethylhydrazobenzene</u>

Edwin M. Kaiser and Gregory J. Bartling, Department of Chemistry,
University of Missouri-Columbia, Columbia, Mo.   65201

$$\emptyset_2CH_2 \xrightarrow[\substack{THF \\ hexane \\ 25°}]{\substack{\underline{n}\text{-BuLi} \\ \underline{i}\text{-Pr}_2NH}} [\emptyset_2CHLi] \xrightarrow[\substack{2.\ NH_4Cl \\ H_2O}]{\substack{1.\ \emptyset N=N\emptyset \\ THF \\ -78°}} \emptyset_2CHN{\overset{\emptyset}{N}}H\emptyset$$
$$93\text{-}98\%$$

1881   S(-)-α-(1-Naphthyl)ethylamine

E. Mohacsi and W. Leimgruber, Chemical Research Department,
Hoffmann-LaRoche Inc., Nutley, N. J.   07110

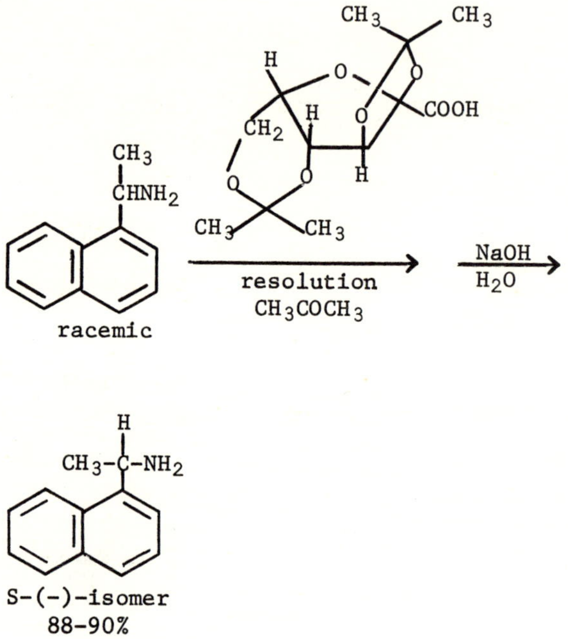

S-(-)-isomer
88-90%

1882   Ethyl γ-Chloroacetoacetate

E. Campaigne and R. B. Rogers, Department of Chemistry, Indiana
University, Bloomington, Ind.   47401

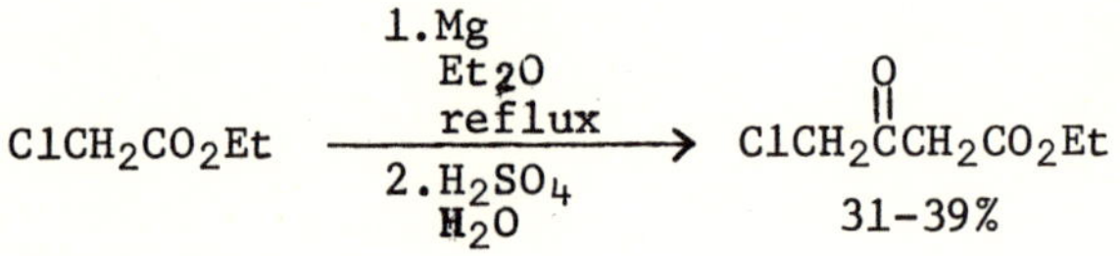

$$ClCH_2CO_2Et \xrightarrow[\substack{2.H_2SO_4 \\ H_2O}]{\substack{1.Mg \\ Et_2O \\ reflux}} ClCH_2\overset{O}{\overset{\|}{C}}CH_2CO_2Et$$

31-39%

1883 <u>trans</u>-4-Hydroxy-2-hexenal <u>via</u> 1,3-Bis(methylthio)allyl-

lithium

Bruce W. Erickson, Department of Chemistry, Rockefeller University, New York, N. Y. 10021

$$CH_3SCH_2\overset{\overset{\textstyle OCH_3}{|}}{C}HCH_2SCH_3 \quad \xrightarrow[\substack{pentane \\ THF \\ -75°-\ 0°}]{\substack{\underline{i}-Pr_2NH_2 \\ \underline{n-BuLi}}}$$

$$\left[(CH_3SCH\text{-----}CH\text{-----}CHSCH_3)^{\ominus}\ Li^{+}\right] \xrightarrow[\substack{2.NH_4Cl \\ H_2O}]{\substack{1.CH_3CH_2CHO \\ -75°\ to\ 20°}}$$

$$CH_3CH_2\overset{\overset{\textstyle SCH_3}{|}}{\underset{\underset{\textstyle OH}{|}}{C}}HCHCH=CHSCH_3 \quad \xrightarrow[\substack{H_2O \\ THF \\ 55°}]{\substack{HgCl_2 \\ CaCO_3}}$$

89%

(4 isomers observed
by 3 peaks in v.p.c.)

$$\underset{\underset{\textstyle OH}{|}}{CH_3CH_2CH}\diagup^{H}\overset{\diagdown^{CHO}}{\underset{\diagup_{H}}{C=C}}$$

41%

(37% overall; pure by v.p.c.)

1884 3-Methyl-2,4-heptanedione

P. Loeliger and E. Flückiger, F. Hoffmann-LaRoche & Co. Ltd.,
CH-4002 Basel, Switzerland

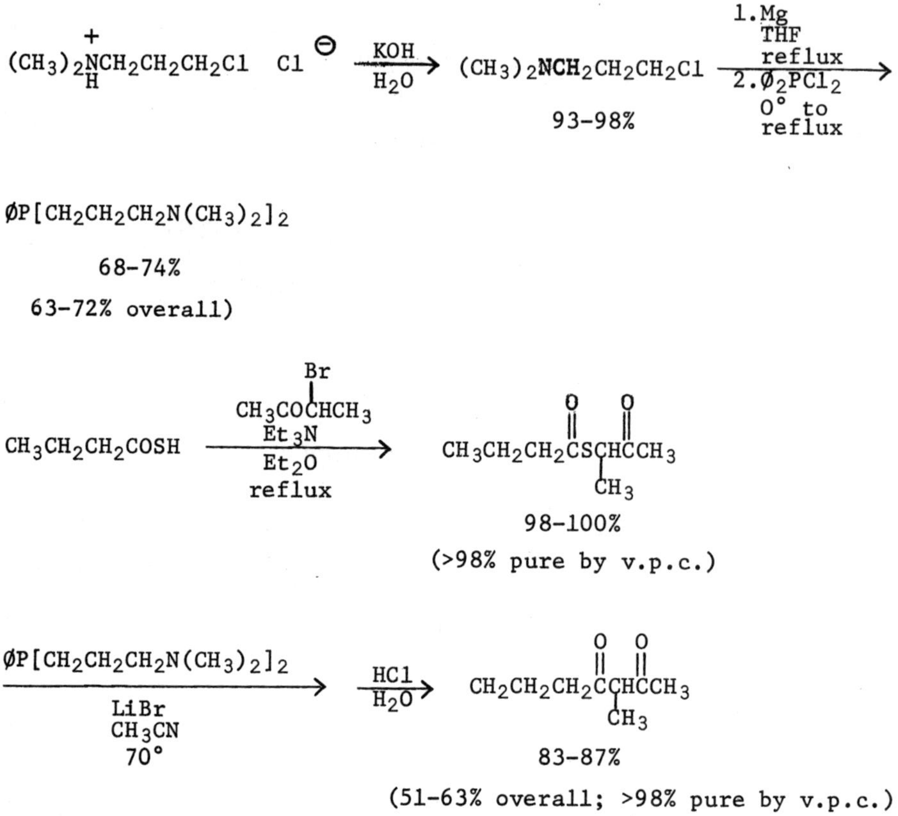

(CH3)2NCH2CH2CH2Cl Cl
KOH
H2O
(CH3)2NCH2CH2CH2Cl
93-98%
1.Mg
THF
reflux
2.Ø2PCl2
0° to
reflux
ØP[CH2CH2CH2N(CH3)2]2
68-74%
63-72% overall)
CH3CH2CH2COSH
Br
CH3COCHCH3
Et3N
Et2O
reflux
CH3CH2CH2CSCHCCH3
CH3
98-100%
(>98% pure by v.p.c.)
ØP[CH2CH2CH2N(CH3)2]2
LiBr
CH3CN
70°
HCl
H2O
CH2CH2CH2CCHCCH3
CH3
83-87%
(51-63% overall; >98% pure by v.p.c.)

1885   <u>N-Methylacetamide Hydrochloride</u>

Richard T. Mayhew and Edward S. Amis, Department of Chemistry,
University of Arkansas, Fayetteville, Ark.   72701

$$2CH_3CONHCH_3 \xrightarrow{HCl} (CH_3CONHCH_3)_2 \cdot HCl$$
$$100\%$$

1886   Preparation of N-Aminoaziridines: <u>trans</u>-1-Amino-2,3-
        <u>diphenylaziridine; 1-Amino-2-phenylaziridine and 1-Amino-</u>
        <u>2-phenylaziridinium Acetate</u>

Robert K. Müller, Renato Joos, Dorothee Felix, Jakob Schreiber,
Claude Wintner, and A. Eschenmoser, Eidg. Technische Hochschule,
Laboratorium für Organische Chemie, 8006 Zurich, Switzerland

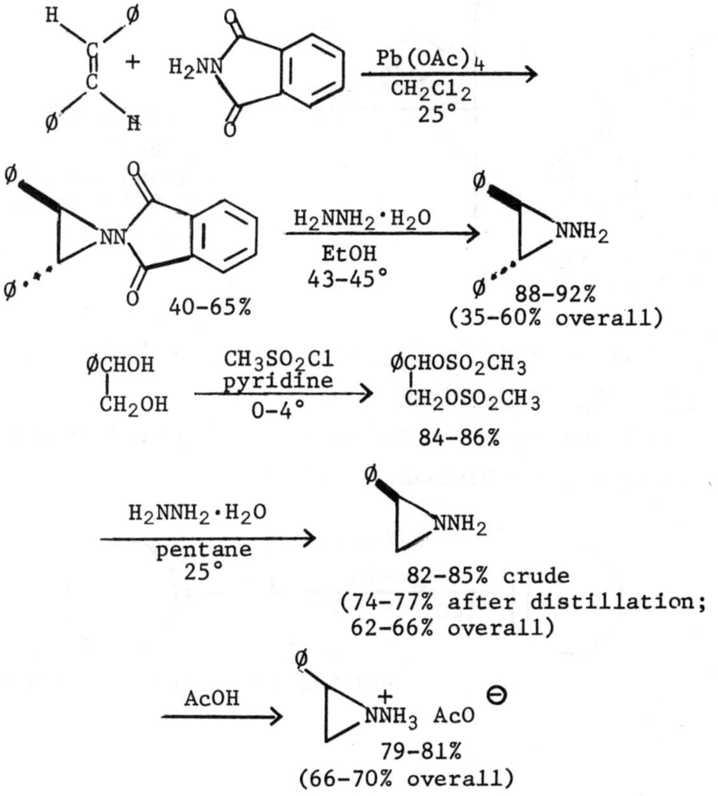

1887   <u>The Fragmentation of α,β-Epoxyketones to Acetylenic</u>

       <u>Aldehydes and Ketones: Preparation of 2,3-Epoxycyclo-</u>

       <u>hexanone and Its Fragmentation to 5-Hexynal</u>

Dorothee Felix, Claude Wintner, and A. Eschenmoser, Eidg. Technische
Hochschule, Laboratorum für Organische Chemie, 8006 Zurich,
Switzerland

71%

diastereomeric
mixture

$HC \equiv C(CH_2)_3CHO$

58-62%
(41-44% overall;
95-97% pure by
v.p.c.)

1888   <u>Hydrogenation of Olefins with a Soluble Lithium/Cobalt</u>

       <u>Catalyst: Cyclooctene</u>

John Carl Falk and James Van Fleet, Roy C. Ingersoll Research Center,
Borg-Warner Corp., Des Plaines, Ill.   60018

cyclohexane   100% conversion

Samir Chandra Lahiri, Jayanta Kumar Gupta, Anindya Mondol, and
Soumendra Mohon Roy, Department of Pharmacy, Jadavpore University,
Calcutta 32, India

$$\text{ØCH=CHCO}_2\text{Et} \xrightarrow[140\ -150°]{\substack{\text{CH}_2(\text{CO}_2\text{Et})_2 \\ \text{Na}}} \begin{array}{c} \text{ØCHCH}_2\text{CO}_2\text{Et} \\ | \\ \text{CH}(\text{CO}_2\text{Et})_2 \end{array}$$

70-72%

$$\xrightarrow[\text{H}_2\text{O}]{\text{HCl}} \underset{87\%}{\text{Ø}\!\!\begin{array}{c}\text{CHCH}_2\text{CO}_2\text{H} \\ \text{CH}_2\text{CO}_2\text{H}\end{array}} \xrightarrow[\substack{100° \\ 3\ \text{hr.}}]{\text{PPA}}$$

CH$_2$CO$_2$H

71-75%
(43-47% overall)